SOTHIS PRESS

Sothis Press aims to completely reimagine the public domain book with an emphasis on universal access to captivatingly designed print editions. We believe that technology will enable every single book ever published to be perpetually accessible in beautiful print form. Sothis is beauty in ubiquity.

www.sothis.press
1 Waterhouse Square, London, EC1N 2ST

Description of the Coal Flora of the Carboniferous Formation in Pennsylvania and Throughout the United States

Leo Lesquereux

SOTHIS PRESS

ISBN 978-1-0365-0772-5

Description of the Coal Flora of the Carboniferous Formation in Pennsylvania and Throughout the United States is in the public domain.

This edition is published by Sothis Press © 2023. The company is registered in England and Wales. The Company Number is 14900389.

Geological Reports of Pennsylvania, 1874–1884.

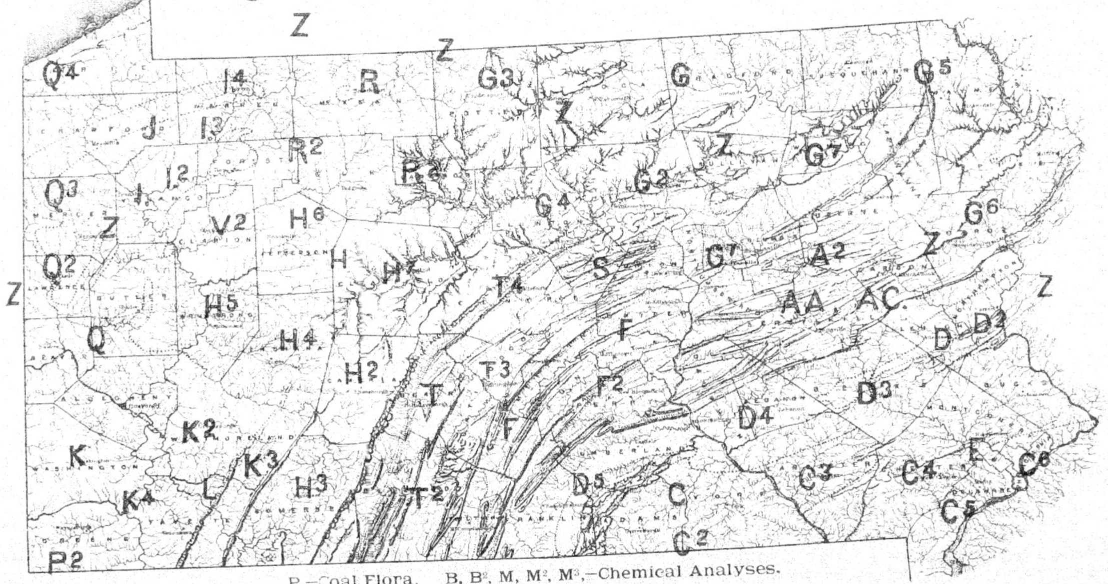

P.—Coal Flora. B, B², M, M², M³,—Chemical Analyses.

N.—Levels. O, O²,—Catalogue of Specimens.

SECOND GEOLOGICAL SURVEY OF PENNSYLVANIA:

REPORT OF PROGRESS.

P.

DESCRIPTION

OF THE

COAL FLORA

OF

THE CARBONIFEROUS FORMATION

IN

PENNSYLVANIA

AND THROUGHOUT

THE UNITED STATES

VOL. III.

BY

LEO LESQUEREUX.

HARRISBURG:
PUBLISHED BY THE BOARD OF COMMISSIONERS
FOR THE SECOND GEOLOGICAL SURVEY.
1884.

Electrotyped and printed by
LANE S. HART, State Printer,
Harrisburg, Pa.

BOARD OF COMMISSIONERS.

His Excellency, ROBERT E. PATTISON, *Governor*,
and *ex-officio* President of the Board, Harrisburg

ARIO PARDEE,	Hazleton.
WILLIAM A. INGHAM,	Philadelphia.
HENRY S. ECKERT,	Reading.
HENRY MCCORMICK,	Harrisburg.
JAMES MACFARLANE,	Towanda.
CHARLES A. MINER,	Wilkes-Barre
JOSEPH WILLCOX,	Media.
Hon. DANIEL J. MORRELL,	Johnstown.
LOUIS W. HALL,	Harrisburg.
SAMUEL Q. BROWN,	Pleasantville.

SECRETARY OF THE BOARD.

WILLIAM A. INGHAM, - - - - - - - - Philadelphia

STATE GEOLOGIST.

PETER LESLEY, - - - - - - - - - - - Philadelphia

ASSISTANTS IN 1883.

Professor I. C. WHITE, geologist, in Huntingdon county; address Morgantown, W. V.

Mr. E. V. D'INVILLIERS, geologist, in Centre county; 711 Walnut street, Philadelphia.

Mr. A. E. LEHMAN, geologist, in Cumberland and York counties; 711 Walnut street, Philadelphia.

Dr. H. MARTYN CHANCE, geologist, in Clearfield county; 2423 Fairmount Avenue, Philadelphia.

Professor E. W. CLAYPOLE, geologist, in Perry and Juniata counties; address in future, Akron, O.

Mr. J. SUTTON WALL, M. E., Monongahela city, Pa.

Mr. A. S. McCREATH, chemist; 223 Market street, Harrisburg.

Mr. LEO LESQUEREUX, fossil botanist; Columbus, Ohio.

Mr. E. B. HARDEN, topographer, in charge of illustrations for reports, and general correspondence at head-quarters; 905 Walnut street, Philadelphia.

Anthracite Survey.

Mr. CHAS. A. ASHBURNER, geologist, in charge of the Survey of the Anthracite coal fields; headquarters, address 907 Walnut street, Philadelphia.

Mr. CHARLES B. SCOTT, assistant and secretary, Philadelphia office.

Mr. O. B. HARDEN, topographer and artist, Philadelphia office.

Mr. FRANK A. HILL, assistant geologist, in the Northern Coal Field; Scranton, Pa.

Mr. JOHN C. BRANNER, topographer, in the Northern Coal Field; Scranton, Pa.

Mr. T. J. WILLIAMS, assistant, in the Northern Coal Field; Scranton, Pa.

Mr. A. D. W. SMITH, aid, in the Northern Coal Field; Scranton Pa.

Mr. ARTHUR WINSLOW, assistant geologist, in the Eastern Middle Coal Field; Philadelphia office.

Mr. WILLIAM GRIFFITH, assistant, in the Eastern Middle Coal Field; Pittston, Pa.

Mr. BARD WELLS, assistant geologist, in the Western Middle Coal Field; Pottsville, Pa.

Mr. H. N. SIMS, assistant, in the Western Middle Coal Field; Pottsville, Pa.

Mr. BAIRD HALBERSTADT, aid, in Western Middle Coal Field; Pottsville, Pa.

LETTER OF TRANSMITTAL.

To His Excellency Governor Robert E. Pattison, *Chairman of the Board of Commissioners of the Second Geological Survey of Pennsylvania:*

Sir: I have the honor to transmit to you the third and last part of the Coal Flora of Pennsylvania and the United States, by Mr. Leo Lesquereux, with additions and corrections of the first and second volumes (published in 1880); a table of species referred to localities; a table of species referred to formations; and a revised index of generic and specific names, referred both to pages and to plates, for all three volumes.

Four years have passed since the publication of the 87 plates of fossil plants in the Atlas to the Coal Flora. During these years, and partly as a consequence of that publication, large quantities of new material have been sent, from all quarters of the United States to the distinguished botanist for investigation. So many new and remarkable forms were discovered in these shipments that he has been able to select enough to fill 26 additional plates with which to illustrate his new species, not only of leaves but flowers and fruits, some of which are singularly beautiful and surprising.

I remain, sir, yours respectfully,

J. P. Lesley.

Philadelphia, *June 18, 1884.*

INTRODUCTION.

One might get an idea of the prodigious activity of the vegetable life during the Carboniferous Period by merely computing the amount of vegetable matter that had to be provided by nature for the composition of a single bed of coal of average thickness. One might also know something about the multiplicity, the diversity, and some of the peculiarities of the coal plants by an examination of the plates and figures of some special work on the vegetable palæontology of the coal, or by merely glancing at a number of specimens of fossil vegetable remains found in strata of Carboniferous age. But the impression thus received would be a mere obscure shadow of the reality; for what is a bed of coal compared to the mass of combustible matter entering into the composition of a vast system of coal beds, of various thickness, like that of the Carboniferous of North America? And in regard to the degree of reality of our acquaintance with the Flora of the coal we have only to remark that this Flora is merely known, as yet, by fragments of different organs, of leaves, of bark of trees marked with peculiar impressions, of fruits, etc.; fragments disseminated and fossilized in beds of shale, or in strata of divers composition intervening in the conformation of the coal measures. Not only, do the fragments not give to the student a full acquaintance with the characters of the plants, but every day new materials are discovered which represent either simple organs or parts of plants unknown to the palæontologist.

Like the vegetation of the present epoch, that of the Carboniferous has been modified and diversified by local circumstances and is therefore differently represented at different localities if separated by a considerable distance. This can easily be seen in comparing the Coal Flora of Pennsylvania and Ohio with that of Illinois, Missouri, Kansas, etc. A certain degree of relation only is recognizable between the plants of strata of the same stage; but a large number of species are only locally found. The differences in the vegetation are still more marked according to the stratigraphical distribution of the measures, or between the plants found in strata of different horizons; and as new coal fields have been recently opened and coal beds worked in Virginia, Tennessee, Alabama, Georgia, etc., at a lower stage than that of the northern basins, a mass of specimens of fossil plants, not yet known in this country, have been discovered and sent from those localities.

As fast as it was possible to work, I have studied the specimens, preparing plates and descriptions for the publication of a supplement to the Coal Flora of the United States which I am now forced to bring to a close, although collectors are still at work and materials continue to come in. I have to leave therefore a large amount of specimens still unexamined; and I foresee that there is left unknown, for future research and study of the history of the vegetation of the coal, an amount of materials, at least as great and as important as that which has already been published.*

I have written the above remarks as a kind of introduction, first, to a brief record of the essential additions recently made by this new volume to the Coal Flora of the U. States†:

---

* Mr. R. D. Lacoe of Pittston, Penna., has directed for years explorations still continued in the more interesting localities of the Coal Fields of North America. He has thus brought up, at great expense, a collection of fossil plants of divers formations, of insects, crustaceans, etc., which is not only by far the largest and most valuable of any in America but which certainly may compare in this specialty with the richest collections of any of the European Museums.

† The Coal Flora of the carboniferous formations in Pennsylvania and throughout the United States, vol. 1 and 2, Harrisburg, 1880. That work is quoted in this volume as the *Coal Flora.*

and secondly, to a detailed description of the species, which are illustrated in twenty new plates.

The additions referred to are:

1st. A remarkable fucoidal plant represented by a large specimen in a perfect state of preservation, on Plate LXXXVIII. A meager fragment of this plant had been figured in Geol. Survey of Penna. 1858, (Pl. XXIII,) mentioned (p. 884) as an Algoid resembling a *Desmarestia*. The plant was examined in England by Prof. Balfour and there figured without either description or name.

2d. On the *Calamarieæ* and *Equisetaceæ* a number of branches bearing fructified spikes referable by authors to *Calamostachys*, *Volkmannia Macrostachya*, etc. Remains of this kind have been extremely rare until now. Those which have been procured and which are figured and described herewith shed some new light on the question of the fructifications of the *Calamarieæ*.

3d. On the ferns, a number of the new species represent remarkable types as yet unknown in the Coal Flora of the U. States. Others, of large size, define more clearly the characters insufficiently observed before on too small fragments, or afford means of referring to one and the same species separate parts which have been described under different names. We, have besides, a number of species of ferns with the fructifications in correlation with the sterile plants. Also a sterile pinna of the remarkable *Sorocladus sagittatus* has been obtained in connection with a fertile part of that plant, thus confirming the reference of those fructifications to ferns.

4th. The *Lycopodiaceæ* are becoming better known;—by the discovery of specimens of very large cones of *Lepidophloios*, which, by the destruction of the sporanges or of their blades, show under them enormous agglomerations of macrospores;—by species of Lycopods, two of them fructified, of very peculiar beautiful types;—by a large cone of *Macrocystis* or *Lepidocystis*, with sporanges attached to the axis, the upper containing macrospores, and by fruiting specimens of some plants of uncertain affinity which I had related to *Cordaites* under the name of *Tæniophyllum*, but which, by their spores, appear to be close relatives of *Isoetes.*

5th. On the *Sigillarieæ*, two remarkably fine new species have been procured, one in Iowa, another in Pennsylvania.

6th. In the fruits, especially, the discoveries have been very valuable. From a large number of specimens, I have been able to put together in juxtaposition upon three plates, different forms of these organs, as they are found, either covered with their *testa* or separated from it as *nuclei.* These numerous kinds of fruits, so peculiar and so different in their forms, sufficiently show the prodigious diversity of the vegetation of the Coal Measures; and prove also how little we yet know of the plants which they represent; for, excepting a few species referable to *Cordaites* the relationship of all the others is unknown.

DESCRIPTIONS.

CELLULAR CRYPTOGAMOUS PLANTS.

MARINE ALGÆ.

DENDROPHYCUS, *Lesqx.*

Root composed of tubulose? flattened filaments, irregularly branching and widely spreading from the base of the rhizoma. Rhizoma cylindrical, simple, long, and thick. Fronds at first top-shaped, or in a compact globular tuft before their development, very large, tree-like, aud many times divided when fully opened; primary and secondary branches thick, somewhat flat on one side, distinctly dichotomous; branchlets pinnately dichotomous; ultimate divisions cylindrical, narrow, pointed.

DENDROPHYCUS DESORII, *Sp. nov., Pl. LXXXVIII, Fig. 1.*

The roots or radicular appendages of this plant appear widely spreading, being seen in profusion upon the shale in the localities where the fronds are found. They are apparently cylindrical or tubulous, though always flattened upon the shale by compression, irregularly dichotomous, often branching at right angles, 3 to 4 m. m. in diameter, of coriaceous or horny texture, shining on the surface.

None of these radicular filaments have been found in connection with the rhizoma which is 1.20 to 1.50 m. long, perfectly cylindrical, 3½ to 4 c. m. in diameter, simple and regular in its whole length with a rough surface. The toughness of its texture is shown in the fact that the specimen found had preserved the cylindrical form in its whole length. The top of the rhizoma, abruptly enlarged into a globular shape resembling a cabbage-head 17 c. m. in diameter, looked, when broken, like a convolute undeveloped frond, with branches densely rolled together into a ball where the divisions or the relative disposition of the branches could not be distinctly observed. The fronds, very large, 1 to 1.25 m. long by 50 c. m. broad, are composed of cylindrical divisions, the primary and secondary ones being thick, the larger 2 c. m. in diameter, flattened on the surface, all gradually smaller from the base to their ends, closely distichous, dichotomous, flexuous, with oblique multiple sub-divisions, the ultimate two ranked, being very closely pinnately distichous, cylindrical, pointed or gradually narrowing from the middle and effaced at the apex. The color and consistence are that of the stone whereupon it is embedded.

As far as I know, this plant has no marked affinity to any species of fossil marine *Algæ*, being only distantly related, perhaps, by the cylindrical form of its sub-divisions, to the genus *Palæophycus* of Hall, and by its oblique dichotomous branches to *Buthotrephis* of the same author. It belongs to a type far more advanced in development than those of the Silurian fucoids. Comparing it to marine plants of the present epoch, Professor Balfour of Scotland finds its affinity to the genus *Desmarestia*, whose fronds are horny, coriaceous, plane or filiform, with distichous divisions. The great size and the mode of growth are however very different and seem to relate the plant to the *Caulerpeæ* or *Syphonaceæ* whose primary stems are strong, rooting, generally horizontal, throwing up erect branches of various shapes, the branches round, cylindrical at base, representing the part which I have described as rhizoma.

Habitat. This plant was first found in the Red shale below Pottsville, the No. XI Mauch Chunk Red shale or Vespertine of Rogers, near the top of the formation or even in the lower part of No. XII, the Conglomerate measures; also abundant in the same formation, in the bluffs of the Susquehanna above Pittston. Though large specimens were found, none of them did show the fronds in a very good state of preservation. It has later been obtained in splendid specimens in a quarry near Davenport, Iowa, in a bed of clay and hardened sand-rock, traversing like a dyke the corniferous limestone overlaid by the Hamilton Group. This location would put the fossil into the Devonian. But the true age of the clay and sand-rock, which appears to be an abnormal deposit into cavitities of the Corniferous, is not positively ascertained.* The specimen figured belongs to the Smithsonian Institution.

VASCULAR CRYPTOGAMOUS PLANTS OR ACROGENS

CALAMARIEÆ.

The plants of this order have a degree of affinity to the *Equisetaceæ* of the present epoch, differing from them by the verticillate leaves, free in their whole length or confluent at their base only, not connate in part of their length into sheaths, and by their fructifications in sporangifer spikes resembling partly those of the Lycopods, partly those of *Equisetum.*

The new materials which I have had the opportunity to examine since the publication of the Coal Flora have added nothing to what is already known of the stems of *Calamites.* But some important data have been procured in regard to the fructifications of *Asterophyllites* which are considered by authors as branches of *Calamites*, and also to the roots of *Calamites*, the stems of *Annularia* and other organs. All these merit consideration in this review, which is made in following the same order as in the first volume of the Coal Flora.

---

*Notes on our local geology No. XI, by Prof. W. H. Barris. Record of the proceedings of the Davenport Academy of Science (1880), p. 163.

ROOTS OF CALAMITES RAMOSUS, *Artis, Pl. XCII, Figs. 1-4.*

In speaking of the roots or rootlets of *Calamites*, I have said that these organs are very rarely observed in the Coal Measures and that I had not been able to find as yet any trace of these remains attached to stems. This blank has been filled by the discovery near Lawrence, Kansas, of five stems of *Calamites* 1½ to 4 c. m. in diameter, bearing, at their round conical base, rootlets seen embedded into the stone as well as the cicatrices of their points of attachment upon the stems. These rootlets are not in bundles but simple, somewhat enlarged at the point of attachment 1½ m. m. thick and gradually decreasing downward, filiform, scarcely 1 m. m. in diameter. They penetrate the stone, passing outward and downward in a curve, from their point of attachment in regular rows a little above the tubercles or leaf scars, more rarely upon the articulations, the point of insertion being marked by a deep round perforation about half as large as the convex tubercles to which they are parallel. They are however irregular in horizontal distances, the space between them varying from 2 to 4 m. m. The whole length of the radicles cannot be seen. By a vertical breaking of the block imbedding the stems of these *Calamites*, the rootlets are exposed for 1 to 3 c. m. from their origin to the points where they penetrate the stone. The scars of the rootlets, as well as the tubercles of the leaves and the articulations, are distinctly seen down to the rounded or subtruncate part of the inverted cones which constitute the base of the stems. The lowest articulations, less than 6 m. m. apart, are gradually more distant up to the eighth one, where, at 7 c. m. from the base, the space between them is 15 m. m. There is no trace of any rhizoma with these specimens which are all isolated. But another specimen of the same locality has three stems, two of which appear to originate from an inclined stock resembling a fragment of *Calamites*, as if they were attached to it like branches, while another gradually narrowed to a point or to a sharply obconical base, bears rootlets and merely touches the stock by its point, in the same way as are represented the young shoots derived from an horizontal branch in Grand 'Eury's "Fl. Carb.," Pl. I, Fig. 2. In all the specimens observed without connection to a parent stock, the basilar point of the stem is a small blunt mamilla, smooth or without any trace of fracture or point of attachment, and the radicles are diverging all around the conical base.

These specimens fully confirm the observations of Grand 'Eury upon the *Calamites* which he considers as originating in the deep mud of swamps, where they germinate from mere filaments, either without any kind of rhizome, or attached to an original stalk or subterranean stem imbedded in the mud.

The French author describes the rootlets as simple or ramose. In the specimens described here they are all simple. He remarks that he could not see distinctly the mode and the point of attachment of the rootlets, and supposed that they had been fixed to the stems just at the joints of the articulations, or immediately above, as he has figured their cicatrice in Pl. I, Fig. 4, l. c. They are rather higher, at least in one of the specimens, producing in their series a ring marked with small perforations above the tubercles or near the middle of the internodes when these are very close. On another specimen they originate just above the articulations at the base of the ribs.

Habitat—The specimens No. 593 are in the cabinet of Mr. R. D. Lacoe, of Pittston, all obtained from Kansas.

CALAMITES RAMIFER, ? *Stur. Pl. XCI, Figs.* 4, 4a.

The first definition of this species was made from an impression, and the ribs, therefore, have been described as furrows. It should be amended as follows :

Stems small, bark thin, articulations and ribs scarcely marked upon the bark, distinct upon the decorticated surface; ribs close, narrow, flat; furrows very narrowly carinate, forming by impression an acute narrow ridge; tubercles rarely distinct, generally small, round, like pinheads, sometimes in double rows above and below the articulations; branches long, nearly at right angles, narrowed at their point of insertion; articulations generally distant; leaves in verticils of 7 to 12, each 12 to 22 m. m. long 2½ to 3 m. m. broad, flat, lanceolate, acute, slightly narrowed from the middle to the connate base.

Though the internodes are sometimes very long, especially upon the branches, they become shorter toward the base of the stems, the articulations upon one of the specimens being only about 1 c. m. distant.

This species has been found at the same locality in very numerous specimens, from large stems 6–12 c. m. in diameter, always flattened, to branches with leaves still attached to them, as in the specimen figured. It has been somewhat difficult to obtain the branches bearing leaves and to distinctly see the whorls of leaves and their mode of attachment. The mode of division of the branches and branchlets, as seen Fig. 4, is that of *Annularia*, not of *Asterophyllites*. It is true that the same mode of division of the branches, nearly at right angles, is remarked upon the figure given by Geinitz. "Verst.," Pl. XVI, Fig. 2, of *Asterophyllites foliosus*, Lind & Hutt, a species referred by Brongniart and European authorities, as a branch, to *Calamites Cistii*. The relation of the American species to that of Europe is also seen in the leaves, which though comparatively broader and shorter than those figured by the German author, have about the same facies. Nevertheless I cannot admit the identity of the American species with that represented by Geinitz, l. c., and still less with that of Lindley, for the following reasons:

1st. The stems of the plant which I refer to, *Calamites ramifer*, positively differ from those of *C. Cistii*, Brgt. & Auct., a species common in the American coal measures, having its range of distribution, generally if not always, above the conglomerate, even high above in the measures, and as said already, *Asterophyllites foliosus*, as represented by Geinitz, is considered a branch of *Calamites Cistii*.

2d. The plant described and figured by Lindley and Hutton can not be the same as that figured by Geinitz and still less as the one represented by the American specimen;

for the English authors describe *Asterophyllites foliosus* as "a tall branching plant, with leaves 8-10 in a whorl, perfectly distinct at their base, a little shorter than the internodes, of a linear lanceolate figure, with a slightly falcate direction. There seems to have been a midrib but this is so imperfectly indicated that nothing certain can be indicated about it." In the figure given by Geinitz of this species, the leaves comparatively shorter, are twice as broad, apparently connate at the base, and, as seen Fig. 3a, the nerve is distinct and prominent. The difference in the characters is still greater in the American plant described here, whose leaves are connate at base, still longer and broader than those figured by Geinitz, not soft, but of hard compact texture, not acuminate, but rather obtusely pointed, with the nerve broad and distinct and the branches and branchlets at right angles, having indeed, in its leaves and branches, the characters *Annularia*. It must be remembered, however, that Geinitz identifies Fig. 4 of his Pl. XVI, a fruit bearing spike, with *Asterophyllites tuberculata*, Ll. & Hutt., Pl. XIV and I believe that the fragments of Lindley, figured under this name, represent the same species as the spikes in Coal Flora, Pl. LXXXIX, Figs. 1 and 2, and these beautiful specimens with others of the same kind not figured, have been found in the sub-conglomerate measures of Pittston, like the fragments referred to *Calamites ramifer*. But there is still a wider discrepancy between the figures given by Geinitz of the fruiting spikes which he considers as fructifications of *Asterophyllites foliosus* and those given by Lindley of *Asterophyllites tuberculata*. It is evident that they do not represent the same species. The spikes of Geinitz are of the genus *Volkmannia*, referable to *Asterophyllites* and perhaps to *Calamites Cistii;* those figured by Lindley are spikes of *Annularia*, as recognized by Schimper and as shown by the beautiful figures given by Geinitz himself of the fructifications of *Annularia longifolia*.

From this I am authorized to believe that the stems which I consider as referable to *Calamites ramifer* are stems of *Annularia;* that therefore species of *Annularia* cannot be considered as herbaceous floating plants; but that, like *Asterophyllites*, they belong to the *Calamites*.

Mr. Lacoe writes me on this subject: "I have upon four parts of a once large stone upwards of 50 verticils of *Annularia longifolia*, the leaves of which vary from 1½ to 3½ c. m. in length, the large ones attached to branches 4 to 5 m. m. in width, flattened. These branches come out from larger ones, or stems 15 to 16 m. m. wide, and upon the same stone, there is a *Calamites* stem which appears to be of the same species and measures 7½ to 8 c. m. in diameter. The articulations are 18 to 20 c. m. apart, one of them with a branch still attached, which, 18 to 20 m. m. in diameter, is broken at its second articulation 35 m. m. from its point of origin. Upon the same fragment of stone, there is another limb or branch 12 to 15 m. m. wide, passing under the main or large stem in the direction of its other articulation, and to this limb is attached one of the small branches with verticils of leaves. I have no doubt that all these fragments of stems of various sizes belong to the same species."

BORNIA ROEM.

BORNIA RADIATA, *Schp.? Pl. XCI, Fig. 5.*

"*Coal Flora I*," *p. 30.*

The specimen, a fragment of stem, has narrow convex ribs, with narrow concave furrows of equal width, less than 1 m. m. broad. The ribs are cut straight at the articulations, not alternate at the joints, the internodes being short, from 2–3 c. m. and the articulations marked by a line or depression about 1 mill. wide. There is no trace of tubercles or leaf-scars above or below the articulations, but at the base of each of the ribs and disconnected from them, there is a small punctiform scar like those I have described above as scars of rootlets. Here they are in regular order, like the tubercles of *Calamites*. The most remarkable part of the specimen is the row of contiguous large round cicatrices resembling branch-scars, placed upon one of the articulations. These cicatricies vary from 5 to 7 m. m. in diameter, are exactly circular, with a deep central point appearing like a vascular scar coming out of the stem a little below the articulation. Except this the surface of the scars is quite smooth. Only the half of one of those round cicatrices is seen at the border of the specimen upon the lower articulation.

These scars cannot be considered as scars of branches, not even of adventive branches. They are also too large for scars of roots or rootlets. Their position in a row and contiguous to each other is abnormal, and as far as I know, nothing of that kind has been observed upon stems of *Calamites* or of *Bornia.* Prof. Stur, who has particularly studied the plants of these genera procured from the Culm of Germany, has figured, "Culm Flora," Pl. II, Fig. 2, a specimen of *Bornia radiata*, which he considers a very finely preserved one, and on which the internodial lines, the scars of rootlets, those of the leaves or tubercles and even those of some branches, are distinctly marked. But these branch-scars, like those represented upon stems of *Calamites ramosus*, *cruciatus*, &c., figured by Brongniart, are distantly scattered and have a diameter of only 4-5 m. m. upon stems much larger than the one I have figured. The fragment described by Stur and quoted above, is also a piece of a large stem. The branch-scars are all at a distance from each other, some small, not larger than the tubercles; others larger, but of irregular outline, not in any way resembling those of the American specimen. Some of the branch-scars in Brongniart's figure of *Calamites cruciatus*, "Veg. Foss.," Pl. 19, may be compared to these in their shape, but they are also of very different size and always at a distance from each other and their surface is never smooth, but marked by grooves disposed star-like from the center to the circumference, like the scars left by detached branches on the trunks of *Calamites ramosus.*

Habitat—The specimen was sent by Prof. F. L. Harvey from Fayetteville, Arkansas.

BORNIA RADIATA ?, *Brgt. Pl. XCIII, Fig. 2*, (*a branch.*)

The figure of this plant represents a branch of which I have given a short description in "Coal Flora," p. 31, second paragraph, referring it to *Bornia radiata*, Brgt.

In comparing that branch with the figure given of a fragment of a branch representing a verticil of leaves copied from Ettingshausen by Schimper, "Pal. Veget." Pl. XXIV, Fig. 2, the characters appear identical. In *Bornia radiata*, however, the leaves are generally somewhat broader, sometimes flexuous and split or bifid, especially near the apex. The splitting of the leaves is not clearly seen on the American specimen, as the branch covered by a leaf of *Cordaites grandifolius* is only seen, somewhat obscurely, through it. The apices of the leaves appear really bifid, in some places but the appearance may be caused by the crossing of the leaves in their superposition. That fine branch, however, cannot be referred to any species of *Asterophyllites*. *A. longifolius* and *A. rigidus* the only ones which could be compared to it, are far different in the characters of their leaves.

The specimen is from the sub-conglomerate formation, Campbell's ledge, Pittston, and most of the fragments of stems obtained of *Bornia radiata* in the coal measures of the continent are from sub-conglomerate measures mostly found in Alabama, Tennessee and Arkansas. One specimen only from Cannelton is apparently referable to it.

CALAMODENDRON, *Brgt. Pl. XCII, Fig. 5.*

The specimen described, "Coal Flora," p. 32, Pl. LXXV, Fig. 16, agrees in its character with the description given of the genus by Grand'Eury and though flattened, seems to concord in some degree with the beautiful figure by which the French author represents a restored stem of *Calamodendron*.

In the description of the genus, Grand'Eury says of the stems of *Calamodendron*, that they are composed of two kinds of alternating radiating bands, the ones of prosenchymatous fibres without parietal marks, the other of scalariform vessels. These alternate bands being continued lengthwise from an alternative of surfaces corresponding to that of the *Calamites*, the articulations are more or less distant, the furrows are somewhat narrowed before joining them, and that is not remarked upon the *Calamites*. Branch scars are formed upon nearly all the joints, and these scars are more or less similar to those of *Calamites cruciatus*. The thickness of bark, increasing in proportion to that of the wood, represents more or less distinctly the superficial striæ.

From this we have to conclude, that for the reference of flattened fossil *Calamariæ* to *Calamodendron*, we have to consider especially the narrowing of the furrows or costæ in joining the articulations, their mode of diverging toward some points, the scars of leaves or branches placed upon the articulations and the thickness of the bark.

The specimen figured "Coal Flora," Pl. LXXV, Fig. 16, evidently presents these characters. We have now still another of exactly the same nature, only much larger, where the characters indicated above are far more distinctly marked. It is a fragment of a calamitoid stem 22 c. m. long. The ribbed part quite flattened, embedded into a piece of black bituminous shale, is 6¼ c. m. broad, bordered on each side by a flat band of hard compact black matter 1½ c. m. in diameter. The articulations upon the stems are very distinct, 2½ to 4 c. m. distant and slightly constricted. The striæ are quite regular in the middle of the internodes and there 1 m. m. wide, as well as the furrows; but toward the articulations, they become all contracted by two or three, diverging toward small round points apparently scars of tubercles or leaves. Besides, this and at regular distances, six or seven of the ribs are bent at the base out of their vertical direction tending toward large round scars of branches marked by mamillæ 2½ to 3 m. m. in diameter, which are placed also upon the articulations in horizontal line with the leaf scars. On the exposed flat surface, the branch scars are at a horizontal distance of 3½ c. m., not in vertical line with those of the upper and lower articulations, but alternating regularly in their relative position, or placed alternately in fours upon each articulation, as in a quincunxial order. In the small specimen figured, Pl. LXXV, the divergence of the ribs toward points or scars placed upon the joints of the articulations could not be represented on account of the narrowness of the ribs; but even upon that small stem, the points marking the scars of leaves, like those of branches, are distinctly seen with the lens. The relative position is, however, modified by variability in the width of the ribbed stems.

The new specimen, obtained like the other from Cannelton, presents another remarkable appearance. The ribbed part representing the woody layer is at one place only, in the middle, covered by a coating of coaly matter representing the bark, its smooth surface being marked by obscure vertical lines corresponding to the costæ of the stem, and the thickness of that piece of bark is not more than ½ m. m. The smooth compact borders of the ribbed stem which are 1½ c. m. broad are not transformed into coal but stony, like the shale, and it does not seem possible to suppose that they represent the bark, unless it is admitted with some authors (Profs. Williamson and Stur among others) that the bark was of a cellular soft tissue, comparatively very thick, loosely adhering to the stem. It is therefore admissible that, in some cases, the lateral part of the bark has been embedded into clay and has thus been petrified along the stem, preserving its whole thickness, while the upper part, vertically compressed upon the stem, has been transformed into coal. Prof. Stur asserts that the shrinkage of cellular tissue, in its transformation into hard coal, is equivalent in bulk to 26–27, so that a woody mass of 26 meters should be reduced by the process of carbonization to 1 meter of coal. It is about the proportion existing between the bark surface ½ m. m. thick and the stony border which measures 1½ c. m., or is 36 times as thick as the coaly bark. Without further pursuing the discussion in this matter I will remark that both the specimens considered here as *Calamodendron* have the character of *Calamites approximatus*, Brgt., and considering these characters, the only ones observable from non-silicified specimens, they should be referable to the genus *Calamites*.

I have seen lately two large specimens of *C. approximatus*, one 33 c. m. long, 16 c. m. broad, flattened to about 2 c. m., with articulations 2½ c. m. distant, all equal, 14 in number, the whole surface being covered by a thin coaly layer or bark not thicker than a strong leaf of paper. The other specimen, got from the same place and much smaller, is 6 to 7 c. m. broad, not so much flattened, its articulations at the same distance as in the last and all equal, with a bark 3 to 4 m. m. in thickness. In admitting the rate of shrinkage of the bark as indicated by Stur that would show the thickness of the bark to have been originally 11–12 c. m., or much thicker than the stem, for it can not be admitted that stems, embedded into clay and transformed into stony matter, have been subjected to the same degree of contraction in their bulk as the bark, the only part which has been reduced to coal.

On this subject, Prof. Williamson, resuming a discussion upon the structure of the *Calamites*, remarks ;* that there was introduced into the bark of these plants, in a very early stage of growth, a second cambium layer which though reminding us of the cork cambium in ordinary exogenous stems, produced no cork, but prosenchymatous cells. In its young state, the bark of the *Calamites* was a very loose cellular parenchyma ; but in the older stems most of this parenchyma became inclosed into the prosenchymatous tissue referred to, which appears to have constituted the greater portion of the matured bark. The sustaining skeleton of the plant, therefore, was a hollow cylinder developed centrifugally on the inner side of an enclosing cambium zone. This cambium zone, which must have had some protective periderm external to it, formed a cortical envelope which was provided by nature as an apparel of support for stems whose woody zone was too thin comparatively to their length ; for the same author has seen, in a mine in England, a stem of blackened *Calamites* exposed 10 metres long, its diameter being 15 c. m. at one end, 10 at the other, both extremities passing out of sight buried in the opposite sides of the mine, the stems being therefore probably much longer. The true diameter of the cylinder of the stem when not compressed was only 10 c. m. at one end, 5½ c. m. at the other.

---

*Address to the Geological section of the British association, 1883.

Trees of this length and size could not stand up without a thick bark enclosing the cylinder.

On the essential differences which separate the genus *Calamodendron* from *Calamites* as represented by *C. approximatus*, there is still a great deal of uncertainty and of diversity in the opinions of the phytopalæontologists.

Genus ASTEROPHYLLITES, *Brgt.*

Calamocladus, Schp.

The specimens of fossil plants referable to this genus are generally found as stems bearing branches and whorls of leaves, like those which have been described "Coal Flora," p. 35, Pl. II, Fig. 3.

The fructifications which as yet have not been satisfactorily described are rarely found in such a state of preservation that their essential characters may be satisfactorily recognized. They are of two kinds or of different organism for the same species, viz: in small globose sporanges or macrospores, placed in the axils of upper leaves of branches, forming a kind of loose terminal spike, *Calamostachys*, Schp., as in Pl. XCIII Fig. 4; or in receptacles containing sporanges, placed in whorls and in successive rows, forming cylindrical, generally narrow spikes covered with imbricate scales at the outside, like those of Pl. XC, Fig. 4, or Pl. XCIII, Fig. 5. These spikes have generally been described as *Volkmannia* or *Bruckmannia*. Their relation to *Asterophyllites* or to *Calamites* has for a long time been hypothetically admitted and lately been proved by the anatomical researches of Renault. The first kind of these fructifications has its structure clearly exposed by the position of the small globular bodies placed in whorls on the axils of leaves which have the same characters as those of the stems. From observations made upon the spike Pl. LXXXIX, Fig. 5, they appear to represent true macrospores.

The structure of the second kind of fructifications (*Volkmannia*) is much more complex. It has been admirably elucidated by the anatomical analysis made by Binney*

* Observations on the structure of *Calamites* and *Calamodendron* in Mem. of the Palæontological Society of London 1867, p. 18, Pl. IV and V.

and the excellent figures given of the structure of the spikes of what he has called *Calamodendron commune*. Some figures of the same organism have been reproduced by Schimper in "Palæont. Veget.," Pl. XXIII, Figs. 5–10, from specimens prepared for the microscope and preserved in the collection of Dr. Hooker, ascribing the species to the *Asterophylliteæ* under the name of *Calamortachys Binneyana*. The description given by Binney of these fructifications is partly recorded as follows: "The spikes contain a double row of sporange receptacles, placed immediately one above the other. The receptacles are formed by scales which proceed from the central axis of the cones, at first at right angles, and then, when reaching the outside, taking a vertical direction, somewhat like the scales of *Lepidostrobus*, but arranged horizontally in rows one above the other through the whole series, not in a spiral direction. The sporanges are of an irregular egg-shape, slightly elongated and are arranged by fours symmetrically around a thorn-like process or spindle coming from the axis. In each of the receptacles, then, there are six of these series of four, arranged radially with regard to the central axis, so altogether there are twenty-four sporanges in every receptacle. Each sporange has a covering composed of a single row of cells which generally shows evidence of some disturbance, so that the original form of the sporanges is not often well displayed. They are filled with numerous round spore-like bodies, some of them having apparently a triradiate appearance and looking as if they had divided into four sporules. These are not unlike the spores seen in *Lepidodendron Brownii*, but are more transparent, not so dark in color and of smaller size."*

I have been able to see the observations of Binney confirmed, as far at least as it was possible to make a comparison without anatomical process, by a number of American specimens, one of them figured Pl. XC, Fig. 4. From what is seen upon these specimens, I may add to the above description; that the row of scales separating the groups of

* Binney loc. cit., p. 18.

sporanges are connate on their borders in their horizontal direction from the axis to the inflexed border, even sometimes to their apex, forming a flat bottom with a vertical undulated or dentate border. Some of the specimens have had the axis destroyed by maceration so that the rows of scales, become not only free from the axis, but may be detached from each other like plates superposed in a pile.

Some more details on the characters of these fructifications are given in the description of the following species; One of them represented by fruiting spikes of both kinds, found with branches and leaves proving their specific reference, to *Asterophyllites* is therefore described in this genus. The others, whose specific relation is not proved, are provisionally considered under the generic names of *Calamostachys* and *Volkmannia*. Both kinds have been indifferently described by Schimper in the genus *Calamostachys*.

ASTEROPHYLLITIS GRACILIS. *Lesqx. Pl. XCIII, Figs. 3–7.*

"Coal Flora," p. 42, Pl. II, Figs. 4-5a.

From numerous and very well preserved specimens, I am able to give some more details concerning the characters of this fine species. The fragment of the stem, Fig. 7, is the largest I have seen of the plant. It is 8 m. m. broad, irregularly costate, like a branch of *Calamites*; the ribs are convex, flattish, continuous, not alternate at the articulations; the furrows are marked in the middle or in the keel by a thin line; the articulations, slightly inflated, nearly 3 c. m. distant, bear, at the base of the costæ, scars of leaves or small tubercles, as in species of *Calamites*, and one of them has two scars of branches 4 m. m. in diameter. The branches derived from this stem, or those of secondary order are 4 m. m. in diameter and the articulations 1½ c. m. distant. The surface covered by a coaly epidermis, a bark of about one fourth of a m. m. in thickness, appears bistriate, an appearance caused by the presence of a narrow ridge in the middle of the furrows or in the intervals between the costæ. The leaves of the secondary stems, like those of their sub-divisions, are extremely narrow, nearly linear or as thick at the apex as at the base, open, nearly at right angles at base, arched upward in the middle, varying from 2 to 2½ m. m. long. Of course the distance of the articulations is shorter in the last sub-divisions of the stems, the whorls of leaves in the ultimate branchlets being only 2 m. m. distant.

The spikes bearing fructifications are of two kinds. 1st, composed of round axillary sporanges as described and figured "Coal Flora," Pl. II, Figs. 5, 5a, and in this volume Pl. XCIII, Fig. 4, and, 2d, in verticils by 4 of small spikes composed of imbricated appressed scaliform bracts, like those of *Volkmannia*, described above and apparently containing sporanges with microspores. These spikes are not only found in numbers, mixed with branches and fragments of fruiting branches, as Fig. 4; but some of them still bear at their base upon the main stem, as in Fig. 5, small rows of leaves of the same size and conformation as those seen upon the sterile branches. There is therefore sufficient evidence of the identity of these spikes with the plants described as *Asterophyllites gracilis*. They have the same appearance and the same structure as those described by Binney under the name of *Calamodendron commune*, though representing a different species.

Habitat—Sub-conglomerate coal measures: Arkansas, Alabama; the specimens figured Pl. XCIII are from Dade, Georgia. Lacoe's collection N. 353.

CALAMOSTACHYS. *Schp.*

Fruiting spikes of *Asterophyllites* composed of small globular tubercles or macrospores, borne singly, without involucre in the axils of verticillate leaves, these being of the same specific characters as those of the sterile stems and branches.

CALAMOSTACHYS LANCEOLATA, *Sp. nov. Pl. XCI, Fig. 1, 2; XCIII, Fig. 1.*

Volkmannia elongata, Roehl, Foss. Fl. Westph. p. 19, Pl. VII, Fig. 1.

Primary stem narrow, thinly, regularly striate, distinctly articulate; fructifications in small round axillary naked sporanges, or macrospores in verticils of leaves around the articulations, forming long narrow cylindrical spikes; leaves twice as long as the verticils, linear-lanceolate.

These fruiting branches, compared to that figured by Rœhl do not show any difference in their characters, except in the contraction of the branches and the more distant articulations of the European plant. The sporanges, also, are less distinctly or not at all marked and the leaves appear slightly broader, at least in comparison with those figured, Pl. XCI. These differences are not of great importance and may result from the degree of decomposition and the flattening by compression of the vegetable fragments. We see about the same kind and degree of disparity between the two American specimens which, though procured from different localities, come from the same stage of the sub-conglomerate coal measures. In these, the length and width of the spikes differ somewhat, and in one of the specimens, the spores, in the axils of the leaves, are less distinct. The leaves, however, are of the same length, 4 m. m. in both specimens but they are lanceolate in one, narrower and appearing filiform in the other, Pl. XCI. As this last has the sporanges apparently in a more advanced stage of ripeness, the bracts are probably contracted by age or dryness and their form is also somewhat obscured by the dark compound of the shale. On the American specimens, the length of the internodes of the stem varies from 1½ to 3 c. m., it is nearly double upon that of Europe though the stem is rather narrower.

This *Calamostachys* probably represents fruiting branches of *Asterophyllites longifolius*. For the shale bearing the branch figured Pl. XCI, Fig. 1, has also a fragment of a stem, same plate, Fig. 2, which I consider referable to *A. longifolius*. The stem is thinly striate, neither contracted nor inflated at the articulations; the leaves in close verticils are narrow, linear, 2–4 c. m. long, generally lineate lengthwise by erosion of the epidermis, the medial nerve being scarcely discernible.

Schimper's reference of the specimen described and figured by Rœhl as *Volkmannia elongata* l. c. to his *Calamostachys typica* "Pal. Veget.," p. 328, Pl. XXIII, is apparently a mistake,as from his figure, *C. typica* evidently represents like *C. Binneyana* spikes bearing sporanges, while Rœhl figures a true *Calamostachys* or a spike bearing macrospores in the axils of the leaves.

Habitat—The specimen Pl. XCI, Fig. 1, is from Arkansas, the other from Dade county, Georgia, both from sub-conglomerate coal. Lacoe's collection, Nos. 558ª and 558ᵇ.

CALAMOSTACHYS OVALIS, *Lesqx., Pl. LXXXIX, Figs. 3, 4.*

Asterophyllites ovalis, Lesqx., " Geol. of Penna." 1858, p. 851, Pl. I' Fig. 2.

Primary stems thicker than in the last species deeply irregularly costate ; articulations at equal distances; fruiting branches thick and long, in verticils around the articulations ; bracts much longer than the internodes, narrowly lanceolate, incurved, sporanges small oval or globose axillary.

This species is much like the former though specifically different. The specimen Fig. 4 is sterile, or its sporanges have been deteriorated and fallen off. The branches are comparatively thick, 5 m. m. in diameter ; the articulations at the same distance, and the leaves, a little more than 1 c. m. long, strongly nerved, are of the same character as those of *Asterophyllites equisetiformis*, Schloth. The articulations of the main stem, like those the branches Fig. 4, are distinctly marked by small round scars which may be the basilar impressions of detached leaves or more probably of both leaves and sporanges.

The other specimen, Fig. 3, merely differs from that of Fig. 4, by the main stem narrower, the branches longer and narrow, the sporanges present and distinct in the axils of the leaves which are slightly shorter. The difference in size merely results from the position of the fragments, either in the upper, or in the lower part of the stems. It agrees with the gradual upward decreasing of the thickness of the stems and branches, as seen upon a large specimen of *Asterophyllites equisetiformis*, mentioned "Coal Flora," p.

35, which, 88 c. m. long, has the main stem 1½ c. m. broad at base and only 2 m. m. near the broken apex. In the fruiting as in the sterile branches of *A. equisetiformis*, the articulations of the main stems are not inflated, but the costæ are slightly so, just above and below the points of insertions of the leaves. The remark upon the fructifications of *Asterophyllites equisetiformis* "Coal Flora," p. 36, confirms the reference of these fructified branches to that species.

Habitat—Dade, Georgia, Fig. 3, No. 852[a]; Fig. 4, No. 852[b] of Lacoe's collection.

CALAMOSTACHYS BREVIFOLIA, *Sp. nov.*, *Pl. LXXXIX, Figs. 5, 5[a].*

Branches thick; articulations obscurely marked by the bracts which are short, slightly longer than the internodes, thick, rigid, lancelote, acuminate; macrospores axillary, very small.

The figure of this fragment is not quite complete. The space between the articulations, 4 m. m., the comparatively broad and short leaves, 6 m. m. long, are all right, but, in some places, the surface of the flattened spike is covered by flakes of coaly matter containing ovules which are true macrospores, as seen by the enlarged Fig. 5[a]. These flakes appear composed of an agglomeration of the spores detached from the crushed part of the spike. As the specimen cannot represent a *Lepidostrobus*, the presence of the spores agglomerated upon its surface indicates a relation of the fructifications of the *Calamarieæ* to those of the *Lycopodoaceæ*, a relation already recognized as probable by Brongniart. This spike much resembles *Volkmannia gracilis*, St. "Fl. d. Forw." II, p. 53, Pl. XV, Fig. 3 and it might be supposed that in this, as in *Volkmannia Binneyana*, the spores were derived from crushed sporanges covered by the upturned base of connate bracts. The size of the spores, true macrospores, contradicts the supposition.

Habitat—Cannelton, sent by Mr. I. F. Mansfield.

VOLKMANNIA, *St.*

The definition of this genus by Sternberg is vague and may apply to every kind of spikes formed by crowded verticillate imbricate leaves. Of the four species which he describes, *Volkmannia disticha* and *V. arborescens* may be merely branches of *Asterophyllites* with crowded verticils of leaves, not organs of fructifications; while *V. polystachia* and *V. gracilis*, represent the genus as I have described it above in considering the fructification of *Asterophyllites*.

VOLKMANNIA CRASSA, *Sp. nov. Pl. XC, Fig. 1.*

Primary stems large, nearly smooth or irregularly obscurely lineate on the surface, indistinctly articulate; fruiting spikes verticillate by four, thick, cylindrical, narrowed to the point of attachment, truncate or obtuse at the apex, articulations of the spikes close, indistinct, covered by erect crowded narrow lanceolate leaves; those of the stem surrounded by longer incurved broader leaves.

The main stem, 12 m. m. at base, has its articulations 3½ c. m. distant, scarcely marked, and that only by the scars of fruiting branches, and of a few leaves which are seen attached on the borders of the stem; the bark is very slightly and irregularly striate lengthwise; the spikes, somewhat oblique, 12 m. m. in diameter, 9 to 10 c. m. long, with articulations 5 m. m. distant, are covered with linear narrow leaves or bracts 1 c. m. long, erect and crowded, laying upon the spikes and obliterating the articulations. By the form of the leaves or bracts, the fruiting branches are related to *Asterophyllites equisetiformis*. The stem nearly smooth and the narrow leaves are against their reference to this species. The leaves also, attached at the base of the spikes to the articulations, are longer and apparently narrower; hence this *Volkmannia* more probably represents the spikes bearing microspores of *Asterophyllites longifolius*. There is, however, in the upper part of the specimen, alongside of the stem, but not connected with it, two long, comparatively broad leaves, lanceolate-acuminata, 4 c. m. long, 3 m. m. broad in the middle, deeply nerved, which might be related to the spike. As no leaves of *Asterophyllites* are comparable to them, that relation is very uncertain.

Habitat—The specimens have been found in the same coal mine of Dade county, Georgia, with the fragment Pl. XCI, Fig. 2. No. 851, of Lacoe's collection.

VOLKMANNIA PRÆLONGA, *Lesqx. Pl. XC, Fig. 2.*

Calamostachys prælongus, Lesqx., "*Coal Flora,*" *p. 59.*

These spikes described first as *Calamostachys*, loc. cit., are evidently referable to *Volkmannia* and very distinct from all the species known of this genus. No other specimen has been obtained and nothing can be added to the description l. c.

VOLKMANNIA FERTILIS, *Sp. nov., Pl. XC, Fig. 4.*

A fragment of a cylindrical spike 10 c. m. long 2½ c. m. broad, composed of a series of verticillate bracts attached to the axis, horizontally directed to the outside, abruptly curved upward on the borders, contiguous in their length, enclosing, in the space between the verticils, oblong sporanges, either attached to the articulations or to the axis in the space between them, and apparently filled with microspores.

The spike has been partly broken and shows the inner texture of a *Volkmannia* as described by Binney for *Calamodendron commune*. The form of the sporange is not distinct, varied as it is by compression and the irregular fracture of the specimen. But the small bladder-like capsules are clearly seen to be sporanges containing microspores; at least the fracture of one of the sporanges shows the inner part filled by minute round bodies like spores. The cohesion of the borders cannot be observed upon the specimen figured, but as said above, it is distinctly seen upon other specimens which, detached from the axis by maceration, leave the verticils disjoined and separated from each other like small plates. The space between the plates of these specimens is empty or without sporanges. The upper border of the bracts is obtusely cut as in the preceding species, as seen on the upper and lower part of the specimen figured.

Habitat—The last specimen is from Archbald Pennsylvania, roof of Coal B, the others without sporanges and with separating plates were found in Stark county, Illinois. All are in the cabinet of Mr. R. D. Lacoe.

MACROSTACHYA, *Schp. Pl. XC, Fig. 3*; *XCI, Fig. 3.*

"Coal Flora," p. 60.

The genus is described from Schimper. I can add nothing to the description. I have, however, figured two cones of this kind. One Pl. XC, Fig. 3, is in its whole state of preservation and shows remarkably well the outside disposition of the scales and the gradual tapering to the base. It seems to have been sessile and attached by a somewhat large base. The other Pl. XCI, Fig. 3, has the upper borders of the bracts mostly destroyed, is rounded at base in narrowing to a somewhat long curved pedicel, thus showing a different mode of attachment. Pl. III, Fig. 20, of the Coal Flora, which is really a *Macrostachya*, partly shows the inner structure in conformity with that of *Volkmannia*. This is evident, not only from the description of Binney, but from the comparison of the fragment Pl. XC, Fig. 4, with that of Pl. III, Fig. 20. From this and considering the two kinds of branch-scars which are marked upon Pl. III, Fig. 14, it might be supposed that *Macrostachya* stems bear two kinds of fruiting spikes; some larger, attached to the broad scars by their large base; some smaller, depending from the small scars by a pedicellate base. The analogy of internal structure of the cones also proves that the spikes of *Macrostachya* like those of *Volkmannia* are fruiting organs of *Calamites* or of *Asterophylytes*, as already remarked by d'Ettingshausen.

MACROSTACHYA LANCEOLATA, *Lesqx.*

In "Geology of Pennsylvania," 1858, p. 852, I have described as *Asterophyllites lanceolatus* a spike which I compared to *Volkmannia major*, Germ. composed of leaves or bracts united half their length and gradually narrowed or lanceolate in the upper part. These spikes are rare; I have seen only fragments of them. They differ from those of *Macrostachya infundibuliformis* as figured Pl. XC, Fig. 3, in their cylindrical and narrower shape, but are quite as long or perhaps longer. The bracts are narrow, linear, lanceolate above, closely imbricating, so that the articulations are obliterated and somewhat indistinct. The best specimen I have seen is a fragment 10 c. m. long, scarcely 1½ c. m. broad, though flattened, obtained at the railroad cut near the brewery above West Pittston. Its generic relation is as yet unknown. It resembles indeed *Volkmannia major* as figured by Germ. "Verst." Pl. XXXII, Fig. 5. But the leaves or bracts are narrower, quite close to each other and closely imbricate. Schimper remarks on these cones, that they resemble long cylindrical strobiles of *Lepidodendron*, but are regularly articulate with shorter and narrower bracts than those of *Lepidostrobus*. He adds that they are gigantic spikes which appear to have originated upon trunks.

ANNULARIA, *Bgt. Prodr.*

"Coal Flora," p. 44.

In the description of the genus *Annularia*, I have remarked on the fructifications, that they are in long cylindrical spikes, with close articulations and narrowly lanceolate bracts, bearing round sporanges in the axils of the leaves or double oval ones, pedicellate, attached to the middle of the internodes. Of this last kind I have seen no specimen in the American Coal Measures. They belong to *Annularia sphenophylloides*, as seen from the beautiful illustrations given by Stertzel of that species.* Those represented by American specimens are in concordance with the descriptions and figures given by Geinitz of spikes of *Annularia longifolia* "Verst." Pl. XVIII, Figs. 8, 9.

---

*Uber die Fruchtähren von *Annularia sphenophylloides*, T. Stertzel in abdruck a. d. Zeitschr. d. Deutschen geol. Gesellschaft (1882).

ANNULARIA TUBERCULATA, *Pl. LXXXIX, Figs. 1, 2.*

Asterophyllites tuberculatus, Brgt. Prodr.; Lindl. & Hutt., "Fossil Fl." I, Pl. XIV.

Stems large calamitoid; branches bearing fruit, narrow, distantly articulate under the spikes; spikes cylindrical, comparatively very long, composed of close verticils of large globose sporanges placed in the axils of short bracts which, curved first downward at base and then upward from above the base, generally cover them entirely; bracts narrowly lanceolate, acuminate, nerved.

The stem Fig. 2 found upon the same slate as Fig. 1 has the same characters and is most probably of the same species. Both stems are distinctly costate, the first, 12 m. m. in diameter, has the articulations 9 c. m. distant, the small less than 5 m. m. in diameter, has the lowest articulations at 3½ c. m. from the base, the others being gradually shorter until they join the spike where they are about 5 m. m. distant, like the verticils of bracts bearing fruits. The figured specimen does not show the position of the sporanges relatively to the stem and the bracts. That is seen upon a much longer spike preserved nearly entire 20 c. m., long, including the pedicel 9 c. m., of which the axis is exposed in its whole length, the upper part being destroyed by maceration or by the breaking of the slate. The articulations are marked by a slightly elevated narrow ridge upon which are seen scars of leaves on slightly elevated points, and just above, there are small round alveoles with a central point of attachment of the sporanges, which correspond or are in line with the costæ of the internodes. The tubercles or sporanges are globose or broadly ovate, 3 m. m. long, 2 broad. The leaves, attached near the base of the sporanges, are first forced downwards under them, and then curving up from below the middle, they cover them and are appressed against them, sometimes, as I have seen it in some specimens, marking the surface by deep transverse obtuse ridges. The internodes have no trace of scars in the middle or between the articulations. The spikes figured by Lindley and Hutton "Foss. Fl.," Pl. XIV, Figs. 1 and 2, are not in as good a state of preservation as those of the American specimens; they are only fragments of the fructified part without pedicel; but it is scarcely possible to doubt that they represent the same kind of organism, even that they belong to the same species. In the specimen Fig. 1 of our plate, the pedicel, evidently calamitoid and probably a branch of the larger stem, Fig. 2, indicates a reference of the fructifications to some kind of *Calamites.* Both these branches, the larger one especially, with distant articulations, appear like branches of *Calamites ramifer*, or of the species which I have described under that name and which has been found in fragments of stems, branches and leaves, in the sub-conglomerate measures of Pittston. Though it may be, the other specimens found with this, have, as described above, the characters of fructifications of *Annularia* and the identity of all these fragments is certain.

Habitat—Railroad cut near brewery above West Pittston, at or near the base of the conglomerate. No. 914 of Lacoe's collection.

ANNULARIA SPHENOPHYLLOIDES, VAR INTERMEDIA, *Lesqx.*

Branches much and horizontally divided; whorls of leaves measuring laterally or on the broadest side from 8 to 30 m. m. composed of 14 to 22 oblanceolate obtuse more or less inflated leaflets; medial nerves obscure.

It is scarcely possible to say if the specimens are referable to a small form of *A. longifolia* or to a large variety of *A. sphenophylloides.* The leaflets are narrow comparatively to their length, gradually and regularly enlarging to the apex which is generally half round, very rarely marked with a minute point. As the leaflets are inflated, the medial nerve is merely marked by a dark-colored line often indistinguishable. In the largest form of *A. sphenophylloides*, the length of the leaflets is 10 to 12 m. m.; in this variety they vary from 4 to 15 m. m., all inflated, especially toward the apex. As seen from the concave impression of the leaflets, their transverse section would be oval, much like that of *A. inflata*, Lesqx.

Habitat—Lawrence, Kansas; Lacoe's cabinet.

ANNULARIA EMERSONI, *Lesqx., Pl. XCII, Figs. 6, 6a.*

"Coal Flora," p. 50.

The only specimen found has been figured. The description is correct.

ANNULARIA CUSPIDATA, *Sp. nov. Pl. XCII, Figs. 7, 7a.*

Whorls small, of 9 oblong abruptly acuminate or cuspidate small leaves, shorter than the internodes; costa thick, terete, hard, excurrent.

A mere branchlet with 8 verticils of leaflets. The verticils are close, the leaves nearly as long as the internodes, open, at right angles to the stem which is strong, thick and evidently solid, as it is preserved in its cylindrical shape. The verticils are gradually shorter upward, the upper one with leaves 2 m. m. long, the lower ones 3 m. m., about all of the same length in the same verticil, oval-oblong, rapidly narrowed and acuminate by a very thick excurring prominent medial nerve. The shape of the leaves is somewhat like that of *A. Dawsoni*, Schp., comparatively broader and shorter, differing essentially by the acuminate apex and the strong nerve. A comparison with the original specimens obtained from Prof. Dawson shows the species to be entirely different.

Habitat—Rushville, Ohio, sub-carboniferous No. 788. Lacoe's collection.

ANNULARIA MINUTA? *Brgt. Pl. XCII, Figs. 8, 8a.*

Plants very small with a main stem comparatively thick, not quite 1 m. m., obscurely striate; branches at right angles; verticils close; leaves in whorls of 8 to 12, spatulate, obtuse or obtusely pointed, 1½ to 2 m. long, ¾ of a m. m. broad in the upper part toward the apex; the medial nerve totally obsolete.

This species is comparable to *A. Emersoni*, from which it differs by the small size of all its parts, the leaves of a different shape comparatively much broader not acuminate but rather obtuse. It may be *A. minuta?* Brgt. as deribed and figured by Wood.* At least the verticils of

"*Trans. Am. Phil. Soc.," vol. xiii, p. 347

leaves are of the same size and form, though much more distant. As the author has not given a detailed description of the species, it is hardly possible to decide if these specimens are referable to the same.

Habitat—Blackville, Monongalia county, West Virginia. Lacoe's cabinet, No. 210.

SPHENOPHYLLUM, *Brgt.*

"*Coal Flora,*" *p. 51.*

SPHENOPHYLLUM LONGIFOLIUM, *Germ., Pl. XCI, Fig. 6.*

"*Coal Flora,*" *p. 53.*

I have figured, as quoted above, a fine specimen which represents the *var. latifolium* of this species. The variety as well as the normal form are very rare in the American carboniferous.

Habitat—The specimen is from the sub-conglomerate measures of Arkansas, communicated by F. L. Harvey of Fayetteville.

SPHENOPHYLLUM SAXIFRAGÆFOLIUM, *Germ., Pl. XCIII, Figs. 9, 9a.*

Leaves in verticils of six to nine; leaflets long, gradually narrowed and cuneiform to the base, sharply, more or less deeply dentate at the apex; primary veins about four, some of them forking above, each division entering one of the teeth.

Of this species which is said to be very common in Europe, I have seen only the fragments figured, which, like the preceding, have been sent from Arkansas by F. L. Harvey. Schimper considers this species a variety of *S. erosum*, Ll. and Hutt., as described in "Coal Flora," p. 55. The difference which seems persistent is essentially marked in the primary nerves not confluent at base.

SPHENOPHYLLUM ANGUSTIFOLIUM, *Germ, Pl. XCIII, Fig. 8.*

S. densifoliatum and S. tennifolium, Font and White, Upper Carb. Fl. pp. 37, 38, Pl. I, Figs. 7, 9.

Leaves narrow and narrowly cuneate to the base, generally in verticils of four or six, deeply dentate at the apex or split once or twice to above the middle; nerves 2–4 at the base, not connivent, either simple or forking once and generally 4–8 at the apex of the dentate leaflets, only 2–4 and entering the teeth of the narrower ones, which are split once; fruiting spikes long, cylindrical, narrow, terminal on slender branches, or shorter and axillary.

The fructified branch figured has a narrow rachis and the articulations scarcely marked. The stem and branches are broader, 2–3 m. m., distinctly deeply striate, with close inflated acutely ridged articulations, 8-12 m. m. distant. The leaves are described by European authors as being verticillate by 6. Though the specimens I have for examination are finely preserved, I cannot see any more than 4 leaves in each whorl. The substance of the leaves is thick or subcoriaceous. The epidermis smooth and thick obliterates the nervation which is rarely distinctly seen.

Schimper describes the fruiting spikes as lateral, pinnately disposed, and remarks in a note, "Pal. Veget.," p. 343, that the spikes are narrower than in *S. saxifragæfolium*, but that they are not terminal as Germar describes them. In proof of this he says he has received from Germar himself a very fine specimen with leaves exactly as in the typical form, but with spikes axillary, longer and especially narrower than those figured by Germar and with much smaller sporanges, and he refers them as identical to the one figured by Germar in "Verst.," Pl. VII, Fig. 5. Now in that Fig. 5, the spikes are not only much smaller, but also much shorter, and the sporanges appear as round tubercles, like the macrospores of *Calamostachys*. They are also figured in that way by Weiss. "Fl. des jung. Ltk. u. d. Rothl.," Pl. XVIII, Fig. 33. As from our fig. Pl. XCII, Fig. 8, the large spikes are positively terminal, as they are also figured by Germar, the small spikes which are lateral and sessile may represent axillary fructifications with macrospores, as in *Calamites gracilis*, while the large ones, which are terminal, possibly bear, like those of *Volkmannia*, microspores enclosed in verticils of larger sporanges. It cannot be supposed that both Schimper and Germar have examined and described specimens of two different kinds of *Sphenophyllum*; for Schimper remarks, as seen above, that those communicated to him by Germar have the leaves of exactly the typical characters.

Habitat—The species is more generally found in the upper strata of the coal measures. The fine specimen figured is from Pomroy Coal, at the horizon of the Pittsburgh Coal. Another, sterile, is from Monongalia Co., W. Virginia, 400 feet above the Waynesburg coal. Lacoe's cabinet No. 768. Upper Coal Measures, Font and White.

SPHENOPHYLLUM TENERRIMUM, *Ett. Mst. Pl. XCII, Figs. 9–10a.*

Stur "Culm Flora," p. 108, Pl. VII.

Stem slender, foliate and ramose; articulations generally prominent; internodes short, somewhat narrowed and costate in the middle; scars of leaves punctiform, placed upon the articulations; leaves verticillate, rarely simple, generally in whorls of 12, forking, sometimes dichotomously divided in three to six laciniæ; verticils of leaves attached to a diaphragm of the stem which is deciduous with the leaves; fructifications at the apex of the branches, spiciform, covering numerous foliate internodes, sporangifer; sporanges compressed pyriform in outline, sessile, in verticils, alternating with the leaves, the sporangifer placed above the verticils of leaves and filling an internode.

The above description with scarcely any modification is that of Stur. It applies exactly to what is represented by our specimens, except for the fructifications which I have not seen and for the mode of attachment of the leaves. The author says that the leaves are attached to a diaphragm of the stem, while our specimen shows the leaves connate at base all around the stem, the short connate part forming a kind of narrow ring, at right angles to the stem. That may be a diaphragm of the stem; but when a number of leaves are detached from the articulations, they remain connate at base, the connate part being about 1 m. m. in width. In the lower part of the branches or stems, the leaves are all divided into *laciniæ* and not connate at base, but in the upper part, they are all simple, erect and somewhat incurved, very narrow, filiform, $\frac{1}{4}$ or $\frac{1}{3}$ of a m. m. in diameter, 5–6 m. m. long or twice as long as the internodes. In the lower part of the stems, however, they are at right angles, looking like bundles of radicles, always forking or dichotomously cut into *laciniæ*, sometimes from the base, sometimes from the middle and above.

Habitat—The specimens are small fragments of branches or stems scarcely 4 c. m. long. They were procured from the black very bituminous shale of Rushville, Ohio, of the Waverly formation, referable to the lower sub-carboniferous measures. They were collected by Leo Lesquereux, Jr., and are preserved in the cabinet of Mr. Lacoe, No. 310. It is from this deposit that Prof. Andrews has obtained all the specimens which he described in vol. 2 of the Geol. State Survey of Ohio. He has given in Geol. Report of Ohio, II, p. 424, Pl. LIII, Figs. 1, 2, a good representation of two fragments of stems with dichotomously divided leaves, considering them as roots. They are evidently referable to the species described above.

EQUISETITES, *Schp.*

Coal Flora, p. 62.

EQUISETITES GRACILIS, *Sp. nov., Pl. XC, Fig. 5.*

The half of a sheath of Equisetites, attached to a costate calamitoid stem by a crenulate narrow border; segments connate at base for one third of their length, then free, long, nearly linear, slightly recurved above, acuminate, marked lengthwise by a nerve or line distinct to the point of cohesion of the segments.

The border base glued to the stem is 2 m. m. broad only, equally crenulate at the lower border or cut in half round short lobes, each corresponding with the base of the *laciniæ* and separated from them by a deep horizontal depression. The sheath, formed by a smooth membrane connecting the segments and bordered by concave sinuses between them, is 1 c. m. broad, and the segments, 33 m. m. long, including the sheath which they cover, are nearly linear, measuring 2 m. m. at base and 1 m. m. at the point where they become free from the sheath. They are then very gradually narrowed upward to the acuminate apex. It is not possible to see what diametral part of the sheath is represented by the fragment. It seems to be about one half, and as eleven of the segments are preserved, their number upon a whole sheath may have been 20 to 24.

The sheath is upon a piece of shale covered with stems and leaves of *Asterophyllites gracilis*. No stems of this species have been found large enough to authorize the supposition that the fragment of sheath could be referable to it. Its name merely relates to the character of the segments.

Habitat—Dade county, Georgia. Lacoe's cabinet, No. 938.

TROCHOPHYLLUM, *Lesqx.*

"Coal Flora," p. 63.

T. LINEARE, *Lesqx., ibid. p. 64, Pl. III, Figs. 24–25b.*

A large number of specimens referable to this species have been obtained since it was first described. The branches bear either flattened leaves or inflated hollow quadrangular organisms, which appear to be sporanges filled with large macrospores. They are described with the *Lycopodiaceæ*.

FILICACEÆ.

NEUROPTERIDEÆ.

It is well known that the remains of fossil plants, especially those of the ferns of the carboniferous are always more or less fragmentary and merely represent fractions which rarely give a good idea of the real characters of the vegetable in its whole. For this reason the identification and also the descriptions of the fossil fragments of ferns are generally very difficult. For the same reason, also, it is always of advantage to have for study a number of specimens of the same species and to represent by figures the parts of the plants, which, different in some degree, may contribute to satisfy the mind about their general characters. That will explain why I have figured again in this volume leaves or fragments of ferns previously figured and described in the "Coal Flora." The number is not great and certainly, better representations of species not yet sufficiently known, are often more valuable to science than the descriptions of new ones.

In the four plates prepared by drawings of specimens of *Neuropterideæ: Neuropteris decipiens, N. Smithii, N. Elrodi, N. aspera*, with *Odontopteris deformata* and *O. æqualis*, are already partly known. The first had not been figured; all the others, four species of *Neuropteris*, four of *Odontopteris*, one *Megalopteris* and one *Tæniopteris* are new. Of the *Odontopteris*, one species was already known in Europe but not yet found on the American continent.

NEUROPTERIS, *Brgt.*

Coal Flora," p. 75.

NEUROPTERIS CARRII, *Sp. nov., Pl. XCIV, Fig. 4-7.*

Pinnules of various sizes and forms according to their position and the size of the pinnæ. 1st. Small; as small as those of Neuropteris tenuifolia, ovate-oblong, obtuse, cordiform and enlarged on the lower side at base, but not auricled, inclining upward, close and contiguous on the borders, Fig. 7. 2d. Larger, or of medium size, also cordate and slightly unequal at base, deltoid in outline, blunt at the apex, Fig. 6. 3d. Cyclopterid large leaflets, broadly oblong or obovate, rounded at the apex, undulate on the borders, Figs. 4 and 5. Nervation flabellate, dichotomous from the base, very distinct, sharply marked; nerves very distant, oblique or erect and parallel at, and toward the base, gradually curving, becoming less distant in ascending, and reaching the borders nearly at right angles, being there very close and often flexuous and reticulate as in species of Dictyopteris.

This fern is very fine; the leaflets are thick or subcoriaceous; the veins become gradually thicker and more distant toward the base. In the small pinnæ, the leaflets are ½ c. m. long, 6 to 7 m. m. broad toward the base. In those of medium size, the pinnules more than 4 c. m. long are 2½ c. m. broad near the base. Of the cyclopterid leaflets 7 to 7½ c. m. long, one is oval-oblong or slightly obovate, 2½ c. m. broad at the base, 4½ c. m. in the upper part, while, the other, part of which is destroyed, is much enlarged at the top, gradually narrowed and cuneiform to the base.

Roehl, in "Foss. Flora," p. 43, Pl. XXI, Fig. 3, has represented under the name of *Cyclopteris orbicularis*, Brgt., a fragment of leaf which, for its form and the disposition of the nerves is very similar to that of the cyclopterid leaflets of this species. The leaf is large, orbicular; the borders are undulate; the nerves, very oblique at base or nearly at right angles near the borders, become there very close, though not flexuous or reticulate in undulations. It may be the same species as that described above, but it is certainly not *Cyclopteris orbicularis* of Brongniart, a species whose large leaflets are entire, not undulate at the borders, the nerves very distant, but proportionally equally so in their whole length, being at the borders 1½ to 2 m. m. distant, while in the American species they are less than the third of 1 m. m. or 36 per c. m. of space. On the figure in Roehl. the veins are exactly one third of 1 m. m. distant. The same author considers *Cyclopteris Germari* St. and *Filicites conchaceus*, Germ. & Kaulf., as synonyms of his species If we are not mistaken in the reference of the American specimens to European species, the two named forms represent the first, *Neuropteris Germari*, very different from C. *orbicularis*, the second, *N. inflata* both distinct from *N. Carrii*.

Habitat—Kingston, Luzerne co., Penna., middle group of coal seams, and Mazon creek, Ills., N. 434 in Lacoe's collection.

NEUROPTERIS OBLONGIFOLIA, *sp. nov. Pl. XCIV, Fig. 3; XCV, Fig. 2.*

Leaflets comparatively long and narrow, linear or oblong-linear, gradually slightly enlarged above the middle and toward the base, there equal on both sides truncate, sub-

cordate or rounded to a short pedicel attached to the upper side of a somewhat broad rachis striate on the borders smooth in the middle; medial nerve thick, flat, percurrent; lateral veins coming out from the midrib in an acute angle of divergence, distant, passing obliquely to the borders, slightly curving and forking twice, once near the base, once above the middle.

The species has no marked affinity to any other of the genus. By its nervation only it is comparable to *N. Rogersi*, Lesqx., except that the leaflets of this last species have no medial nerve. The leaves vary from 6½ to 10 c. m. long and from 1 to 2 c. m. broad near their base. The lateral veins having about the same degree of divergence at their base as in their upper part near the borders, are there 14 to 16 in the space of 1 c. m.

Habitat—Cannelton, Penna., very rare. The specimens in Lacoe's collection bear No. 951.

NEUROPTERIS DECIPIENS, *Lesqx, Pl. XCIV, Figs. 1, 2.*

"Coal Flora," p. 93.

The description of the species, l. c., is right and agrees with the specimens figured here. The primary pinnæ, very rarely found, are large; the secondary are simple leaflets or trifoliate, composed of one very large oblong or lanceolate obtusely pointed pinnule, with two basilar round or oval ones, attached at its base in the same manner as in *N. hirsuta, N. angustifolia*, etc. The fine specimen Fig. 1 is the first I have seen with pinnæ and pinnules attached to the rachis, these being generally found isolated. They are always larger, in comparison, than those of *N. hirsuta.* I stated in the description, l. c., that counted upon the borders, the nerves in *N. decipiens* are 28 per c. m. I should have said 25 to 28, while in *N. hirsuta* they count 40 to 50 per c. m. The two specimens figured show the great variability in form and size of the pinnæ and pinnules. As in *N. hirsuta*, the leaves, though generally hirsute, are sometimes without hairs.

Habitat—The *habitat* is exceptional. The species abounds at Mazon creek, but as yet I have seen one specimen only from a different locality—the shaft at Centralia, Ills.

NEUROPTERIS ANGUSTIFOLIA, *Brgt.*

Lesqx. "Coal Flora," p. 89, Pl. VIII, Figs. 2, 3, 6, 8, 10, 11.

This species has always been found until recently with the leaflets smooth on the surface. A variety has now been sent from Ottawa and Osage City, Kansas, with leaflets hirsute, like those of *N. hirsuta.* In the form and areolation, the leaves have the same character as those of *N. angustifolia;* they are oblong or linear-lanceolate, comparatively narrow, with a broad deep percurrent medial nerve; the lateral nerves are oblique, somewhat thick, 35 to 40 per c. m. counted along the border, or less than in *N. hirsuta* and more than in *N. decipiens.* With the narrow leaflets, the specimens from Osage City have two round cyclopterid ones deeply auriculate at base, 5½ c. m. in transverse diameter, 4½ vertically or from the upper border to the base of the auricles, which descend 2 c. m. lower than the point of attachment. Their surface appears covered with thin hairs, but as the specimens are varnished, the hairs could not be distinctly seen. It is however most probable that these cyclopterid leaves belong to the same species. But other kinds of Neuropteris have been found at the same locality.

NEUROPTERIS SMITHII, *Lesqx. Pl. XCVI, Figs. 3–3a.*

"Coal Flora," p. 106, Pl. XIII, Figs. 1-3.

The first specimens received of this species from the Black creek vein of Alabama were in large pieces of shale covered only with loose scattered very small pinnules of the plant and a few fragments of pinnæ which have been figured. The description of the species could not be definite nor as detailed as desirable. Other specimens, one of which is figured above, show some difference which it is advisable to record. The rachis is not always very narrow. It is sometime, comparatively broad for the size of the pinnæ. A fragment of a primary stalk on the right side of the specimen, is 1 c. m. in diameter, its surface is rough or coarsely rugose, while the secondary rachis, half a c. m. broad, is flat, distinctly finely striate lengthwise. The secondary and ultimate pinnæ are generally at right angles, linear, and the pinnules are not always distant round and cordate at base, but the upper ones are attached to the rachis by their whole base, decurring, the lower ones only being cordate at base and attached by the middle, or the base of the medial nerve which is more or less effaced in the upper part of the leaflets, though often discernible to the apex. The lateral nerves are more or less arched according to the size of the pinnules and reach the borders rather obliquely than at right angles, generally forking twice. In the upper decurring leaflets, a few of the lateral nerves are derived from the rachis, but in the lower pinnules, they are all from the middle nerve. That kind of nervation and the disposition of the upper pinnules is generally the same in the species of small *Neuropteris*, like *N. tenuifolia*, *N. Desorii*, even, though less evident, in *N. Loschii*, whose leaflets are generally larger.

Habitat— Sub-conglomerate coal measures, Alabama, Georgia, Tennessee, Virginia. I have not seen it yet from Arkansas where *N. tenuifolia*, the nearest relative to the species is present. The specimen figured is No. 543 of Mr. Lacoe's cabinet.

NEUROPTERIS ELRODI, *Lesqx. Pl. XCVI, Fig. 1, 2.*

"Coal Flora," p. 107, Pl. XIII, Fig. 4.

Though the species was first described from a single small fragment of a pinna, its characters are rightly defined. From the examination of a large number of specimens of that fine species more recently obtained, generally from the localities where *N. Smithii* is also found, it is evident that these two species are distinct though related in some points. In comparing the figures of Pl. XCVI, the points of difference are easily remarked. In *N. Elrodi*, the ultimate pinnæ are long, somewhat flexuous, attached to a comparatively narrow rachis ; the pinnules are lanceolate, obtusely pointed, more than twice as long, though not broader ; the ultimate pinnule, especially, is totally different, much longer than the lateral ones, narrowly lanceolate, acuminate and blunt at the apex. The nervation is clear, distinct, of the same character as that of *N. Smithii*. But in some specimens it is rendered obsolete by a thick epidermis which is somewhat rough or villous, appearing as covered with soft thin hairs. The main rachis is smooth not striate.

I have remarked in the description, that the species is closely related to *N. Dluhoschi*, Stur. I now believe that the species are identical. Considering the form of the leaflets, their thick substance, the narrow rachis and the nervation, it is also probable that the fragments described and figured "Coal Flora," p. 121, Pl. XIII, Fig. 7, as *N. biformis*, represent a large form of the same species. The base of the terminal pinnule preserved upon the specimen seems to indicate by its mode of attachment an elongated lanceolate shape, like that of the terminal leaflets of Fig. 1. If this character was ascertained, identification would be positive.

Habitat—Dade, Georgia, and also Tennesee, with the preceding. The specimens figured are in Mr. Lacoe's cabinet, No. 865.

NEUROPTERIS ASPERA, *Lesqx., Pl. XCVI, Figs. 4-7.*

"Coal Flora," p. 121, Pl., XIII Figs. 10-12.

Though the new specimens of this peculiar species do not sufficiently show its real character, they at least prove that the remark made p. 122 of the Coal Flora upon two specimens whose leaflets had the surface hairy, are not to be considered in relation with this plant. They represent a part of a pinna with very oblique pinnules and a terminal leaflet remarkable for its length, enlarged and obtusely lobate on one side in the middle, gradually acuminate to the apex. All the lateral leaflets are very unequal at the base, prolonged at the lower side into a rounded auricle and obliquely truncate or rounded at the upper. The medial nerve is thick up to the middle, narrower and somewhat obscure from the middle upward, divided at the top, at least in some of the leaves, into separate nervilles which fringe the apex like bristles. The lateral veins, very close and at the same time prominent, give to the surface of the leaflets a peculiarly rough appearance by which even the smallest fragments can be identified.

Habitat—Very rare. First obtained at Murphys borough, Ill., later at Cannelton and still later at the Campbell's Ledge of Pittston, Pa. No. 935 in Lacoe's collection.

NEUROPTERIS BLISSII, *Sp. nov.*, *Pl. XCV*, *Figs. 1*, *1a.*

Pinnæ linear-oblong, simply pinnate; secondary pinnæ or pinnules open, nearly at right angles, parallel, alternate, distant, not contiguous on the borders, oblong-linear, narrowed into an obtuse apex, symmetrical and rounded at the scarcely cordate base; medial nerve, thin, effaced near the apex; lateral veins dichotomous, arched, all derived from the medial nerve, distinct, somewhat distant; rachis narrow.

The pinnules average 5 to 6 c. m. in length, the upper gradually shorter, 10 to 15 m. m. in width, being a little more enlarged at the rounded base, attached to the rachis by the base of the medial nerve and more or less undulate on the borders.

The species is related to *N. Volzii*, Brgt , "Veget. Foss." p. 232, Pl. LXVII, a species of the Trias of France, which has, however, a much stronger round rachis, the pinnules more narrow, deeply cordate and auriculate at base, attached to the rachis, says the author, by the medial nerve and the upper side of the base, the lateral nerves being thinner and closer. By the form and size of the pinnules, this new species has also a degree of likeness to *N. angustifolia*, Brgt., differing by the pinnules less enlarged toward the base, linear, all equal, nearly at right angles to the rachis and the lateral veins more distant, generally forking twice.

Habitat—Mazon creek. Communicated by Mr. Frank T. Bliss.

NEUROPTERIS GRIFFITHII, *sp. nov.*, *Pl. XCV*, *Figs. 3–8.*

Separate pinnules of various sizes and forms, generally small, unequally, deeply cordate and auricled at base, either longer, narrowly lanceolate, blunt at the apex, or shorter and broader, oblong, obtuse, or still more enlarged, nearly round or reniform, even narrowed and wedge-form at base, entire or sometimes divided into irregular obtuse lobes; nerves dichotomous, flabellate from the base or derived from a thick percurrent medial nerve, slightly arched, close and distinct.

The pinnules have all been found separated from the rachis. They are all comparatively small, 3 to 5 c. m. in length, 1 to 2 c. m. in width near the base, deeply unequilateral at the auricled base, one of the auricles being prolonged downward and covering or surrounding the point of attachment as in Figs. 6–8, or merely cordate at base, apparently reniform and cyclopterid, as in Fig. 5. In Fig. 3, the leaflet enlarged around a broad base is divided into irregular lobes, like the leaves of *Odontopteris monstruosa*, described below, and in Fig. 4, the pinnule is enlarged upward from a cuneate base.

This last leaflet bears near the borders small tubercles or sori (?) placed in a regular row parallel to the borders, each upon a branch of the nerves. They are very small, not more than half 1 m. m. long and half as broad, apparently convex, as seen Fig. 4*a* enlarged and of the same nature as those which have been observed upon leaflets of *N. Clarksoni*, as mentioned "Coal Flora," p. 95, and already figured "Geol. of Penna." 1858, Pl. V, Fig. 3. By their size, their form and the regularity of their position, these tubercles resemble the sori discovered on the borders of leaflets of *Odontopteris sorifera* Grd'E., mentioned "Coal Flora," p. 124, which are evidently organs of fructifications. As the tubercles formerly observed upon leaves of *Neuropteris*, are more or less irregularly placed in relation to each others and also to the veins, generally near the medial nerve, rather between than upon the veins, their nature was less distinctly indicated than it is upon that leaflet. I have, however, not been able to see the granulations of surface figured by Grand'Eury, marking the place and the disposition of the sporanges in parallel rows.

Habitat—Port Griffith, E vein, near Pittston, Pa. No. 417 Lacoe's collection.

Odontopteris, *Brgt.*

"Coal Flora, p. 124."

As an addition to what has been already published of the peculiar plants of this genus, I have represented upon Pl. XCVII of this volume six different species, two of which, *O. monstruosa* and *O. patens* are new; another, *O. neuropteroides* Roem., is new for America, but was already known in Europe; a fourth, *O. affinis*, formerly mentioned without name "Coal Flora," p. 135, was neither figured nor described; a fifth, *O. deformata*, already described had not been figured in the Coal Flora, and the sixth, a small pinna of *O. æqualis*, "Coal Flora," p. 135, represents the species in a different form, if not with a different character.

In the description of the genus l. c., considering the great affinity of its species to those of *Neuropteris*, I have remarked, p. 124, that as far as observed until now, the more marked difference between these genera was the absence of large cyclopterid leaflets in species of *Odontopteris*, while pinnules of that kind are generally found in connection with pinnæ or fragments of *Neuropteris*. That assertion seems contradicted by the shape and size of the leaves of one of the new species above mentioned, *O. monstruosa*. As seen Pl. XCVII, Figs. 1, 2, that species is represented upon two of the specimens by very large lateral pinnules, oblong or nearly round, with distant flabellate dichotomous nerves, resembling indeed the cyclopterid leaflets of some *Neuropteris*. The likeness is however fictitious, for the conformation of the two kinds of pinnules is very different. Those of *Odontopteris*, in the figures quoted above, are lateral, attached to the rachis by the whole length of a long base. They are evidently formed by cohesion of the borders of a number of lateral leaflets, from which is produced a compound lamina. The different parts, if not distinct, are indicated by the indentation of the borders or by lobes. That is clearly seen on the right side of Fig. 2 and more distinctly on the left of Fig. 1, these being partly connate. The true cyclopterid leaves of *Neuropteris* are formed by the normal expansion of the blade or leaf around a central point, the base of a rachis, which they surround. It is a natural unfolding of the lamina, resulting from the position of the organ, not a deformation like those observed and so very much varied in many species of *Odontopteris*. In the small specimen of *O. alpina*, "Coal Flora," Pl. XIX, Fig. 4, the deformation is of the same kind as in *O. monstruosa*.

ODONTOPTERIS NEUROPTEROIDES, *Roemer*, *Pl. XCVII, Fig. 8.*

Roemer, "*Beitr. in Paleontogr*," *IX, p. 187, Pl. XXX, Figs. 2a, 2b.*

Pinnæ narrowly lanceolate; pinnules disjoined to the base, somewhat distant, inclined upward, oblong-lanceolate, obtuse or blunt at the apex, attached to the rachis by the whole decurring base; terminal pinnule long, rhomboidal-lanceolate, blunt at the apex; medial nerve scarcely distinct; the lateral very oblique, nearly straight in passing to the borders.

The figure of Roemer represents a pinna with pinnules broadly rounded at the apex and a medial nerve very distinct. The description says, however, that they are blunt at the apex (*stumpf*) and that the medial nerve is not or scarcely marked (*undeutlich*). The description is therefore concordant with the characters of the small and only specimen we have of that species. That of the terminal pinnule is translated from the author's. The lateral pinnules are about 2 c. m. long and 5 to 7 m. m. broad in the middle. The lower lateral nerves are derived from the rachis, the others from the medial nerve, all in a very acute angle of divergence.

Habitat—Rhode Island, No. 430 in Mr. Lacoe's collection.

ODONTOPTERIS PATENS, *sp. nov.*, *Pl. XCVII, Fig. 7.*

Pinnæ linear or narrowly lanceolate, long; pinnules nearly at right angles to the broad rachis, narrowly linear-lanceolate, enlarged and connate at base, acute or blunt at the apex, distant; medial nerve thin but distinct to near the apex; lateral veins at an acute angle of divergence, slightly curved in passing to the borders, thin, the lowest only emerging from the rachis.

This species is related to the preceding, from which it differs by the very broad rachis, 5 m. m. in diameter, slightly punctulate-rugulose, the form of the pinnules nearly at right angles, not decurring, but abruptly enlarged at the base and there connate, the form of the narrow leaflets 2 c. m. long, scarcely 5 m. m. broad in the middle, slightly enlarged below the apex and narrowed into a more acute point.

Habitat—Rhode Island. The specimen No. 890 of Lacoe's collection represents three fragments of a pinna of the species upon a large plate of shale partly covered with branches of *O. Brardii.*

ODONTOPERIS MONSTRUOSA, *Sp. nov., Pl. XCVII, Figs. 1–3.*

Fragments of multiform pinnæ, either pinnately divided into obtuse alternate decurring large leaflets, or bearing, on one side of a narrow rachis, a row of oblong obtuse leaflets connate at base and decurring, and on the other a cyclopterid leaf merely lobate on the borders; nerves dichotomous, parallel at base, diverging and flabellate upward, slightly curved or nearly straight, all derived from the rachis.

The figures give an exact representation of these fragments, which it is scarcely possible to clearly describe. Fig. 3 shows the upper part of a pinna with 4 pairs of alternate oval-oblong obtuse leaflets, the lower 2½ c. m. long, 14 m. m. broad, narrowed and rounded above the decurring base, the upper narrower, attached to the rachis by their whole decurring base, the terminal broadly ovate. In all the lateral leaflets, the nerves are derived from the rachis, all dichotomous in ascending, slightly curved toward the borders. In Fig. 1, the flexuous rachis bears on one side 4 leaflets, connate at base, oblong, obtuse, gradually shorter upward, nerved as in Fig. 3, and on the other side, the base of a large cyclopterid leaf which seems attached to the whole length of the rachis, being probably lobate on the borders, as on the right side of Fig. 2. In this last, the rachis also flexuous, has cyclopterid leaflets on both sides, one nearly round, obtusely lobate on the borders, with two narrow obtuse leaflets above, the other entire to the truncate apex, with the base of a narrow terminal leaflet at the top of the medial nerve. The cyclopterid parts have the nerves more distant, but the character of the nervation is of the same type, as seen in Fig. 1, where both small and large leaflets are upon the same rachis. For this reason and considering the extraordinary variety of forms seen upon species of this genus, as for example upon *Odontopteris heterophylla*, Lesqx., "Coal Flora," Pl. XXII, Fig. 6, I refer all these fragments to the same species.

Habitat—Cannelton; found also in nodules at Mazon creek, Ills., No. 756 in Lacoe's collection.

ODONTOPTERIS AFFINIS, *sp. nov., Pl. XCVII, Fig. 4.*

Pinnæ linear-lanceolate; pinnules sub-opposite, large, oblong and enlarged above, attached to the rachis by the whole base, slightly decurring, close to each other, connate at base, the lower pair at right angles, the others gradually more oblique, turned upward; nerves close, all derived from the rachis, parallel, forking once above the middle, straight, or inclined toward the borders.

The lower pinnules are 2 to 2½ c. m. long, 1½ c. m. broad above the middle, the upper ones, gradually shorter and narrower, are disconnected to the base, except near the apex of the pinnæ, where they are connate to the middle, enlarged above, rounded at the apex, with nearly the same form as those of *O. subcuneata*, "Coal Flora," Pl. XXII, Figs. 4 and 5. They merely differ by the entire, decurring, not auricled base, the leaflets more enlarged upward, the nervation a little more distinct and distant, the nerves forking once only. I have not seen a specimen with a terminal pinnule. From the fragment left at the top of the pinna, Fig. 4, that pinnule seems to have been enlarged and short. Another specimen, not figured, has all the leaflets connate to the middle, with a facies quite different from that of *O. subcuneata*. Nevertheless this may be a mere variety of the same species.

Habitat—Mazon creek. Rare.

ODONTOPTERIS ÆQUALIS, *Lesqx.*, *Pl. XCVII, Fig. 6*.

"Coal Flora," p. 135, Pl. XXI, Fig. 8.

A small pinna, evidently referable to this species by its inflated pinnules and the nervation, differing only by the pinnules close, oblong, obtuse, larger at the decurring and connate base. The rachis is thick, 3 m. m. at base, the pinna is only 5 c. m. long, the pinnules, 8 pairs, gradually shorter upward, the leaflets less than 1 c. m., the veins parallel and distant. The pinna has, taken altogether, the same characters as that of Pl. XXI, Fig. 8, merely modified by its different position upon the frond.

Habitat—Mazon creek, in concretions.

ODONTOPTERIS DEFORMATA, *Lesqx.*, *Pl. XCVI, Fig. 5.*

"Coal Flora," p. 141.

The description of the species in the Coal Flora does not conform with the figure of this plate, for the reason that it was made from a different specimen and that all the specimens of this species represent it in a different way. As seen on the figure, the lateral pinnæ of the left side are more elongated and proportionately narrower and the lower pinnules shorter and broader, while on the other side, and by contraction, the pinnæ are shorter and broader, the terminal pinnule, also, much shorter and broader, while the lower pinnules are longer and narrower.

The specimen is a remarkable exemplification of the dimorphism caused by expansion in the lateral compression of a stratum of shale, folded into short undulations along the beach near Newport, R. I., where all the plants have been more or less deformed by traction to one side or to the other according to their position in the bed

TÆNIOPTERIS, *Brgt.*

"Coal Flora," p. 153.

TÆNIOPTERIS TRUNCATA, *sp. nov.*, *Pl. XCIV, Fig. 8.*

Frond pinnate; pinnæ at right angles to a narrow rachis, simple, linear, attached by a rigid percurrent medial nerve, obliquely truncate on both sides at base, somewhat

undulate on the borders; lateral nerves at a broad angle of divergence, but not at right angles to the medial nerve, slightly arched in traversing the blade, extremely thin and close.

The reference of this fragment to *Tæniopteris* is not quite certain, on account of the peculiar nervation, but I do not find it related to any other genus, and it would not be proper to constitute a genus with a mere fragment like this. It has a degree of likeness, by the form and size of the pinna, to *T. coriacea*, Goepp., "Perm. Fl." p. 130, Pl. VIII, Fig. 4 and IX, Fig. 2, differing especially in the narrow medial nerve, the close thin lateral veins and the leaf truncate at the base. The veins are so thin and close that they are perceived with difficulty, even with a strong lens, appearing here and there as undulate and mixed. The fragment of pinna is 4 c. m. long, 1 c. m. in diameter and broken at the apex, which was apparently obtuse.

Habitat—Sub-conglomerate Campbell's Ledge, Pittston, Penna., No. 860 of Mr. Lacoe's cabinet.

MEGALOPTERIS, *Daws.*

"Coal Flora," p. 147.

MEGALOPTERIS RECTINERVIS, *Sp. nov., Pl. XCVII, Fig. 9.*

Primary and secondary rachis narrow; leaflets very oblique, crowded, narrowed at the upper side toward the base, and decurring at the lower, long, linear-lanceolate; veins at a very acute angle of divergence, not more than 10°, thin and close, forking once or simple.

Of this species, there is only the small fragment figured. It shows part of three leaflets, none of them entire, about 1½ c. m. broad, linear, probably narrowed above and obtuse, narrowed on the upper border which joins the medial nerve at its base, long and broadly decurring on the lower side, about in the same way as in *M. Southwellii*, Pl. XXIV, Fig. 1. It differs greatly from this and all the other species known of this genus by the narrow leaflets and the nearly straight lateral nerves.

Habitat—The fragment was sent from Arkansas by Prof. F. L. Harvey.

ALETHOPTERIDEÆ.

"Coal Flora," p. 162.

This group has not been defined l. c. Its species are generally in large fronds or pinnæ, pinnately divided and sub-divided. The rachis is strong; the pinnæ open, large, parallel; the pinnules attached to the rachis by their whole generally connate base, are open or at right angles to the rachis, linear or oblong, obtuse, rarely acute, mostly entire; the medial nerve is thick from the base to the middle, often percurrent, the lateral nerves distinct, very close and at right angles, or less crowded, oblique and curved backward in the genus *Callipteridium.*

CALLIPTERIDIUM, *Weiss.*

"Coal Flora," p. 164.

CALLIPTERIDIUM INÆQUALE, *Lesqx., Pl. XCVIII, Fig. 1.*

"Coal Flora," p. 168, Pl. XXXIII, Fig. 2-5.

All the specimens represented l. c. are fragments of pinnæ of the third order. That of Pl. XCVIII is the upper or terminal part of a secondary pinna with short lateral pinnæ, the lowest deeply pinnately lobed, the lobes gradually shorter in the upper pinnæ which become simple lanceolate obtuse pinnules like the terminal one. This part shows the general character and aspect of the plant. The texture is thick and the nervation very close, obsolete, as it appears also upon most of the smaller specimens of the species, all obtained at the same locality, Cannelton, Penna.

CALLIPTERIDIUM SINUATUM? *Pl. XCIX, Fig. 3.*

Pecopteris sinuata, Brgt., Veg. Foss., p. 296, Pl. XCIII, Fig. 3.

Leaflet oblong-lanceolate, gradually narrowed to an obtuse apex, more prolonged on one side of the unequilaterally cordate base; surface obliquely sinuous, plaited, but not laciniate or cut through by the lateral nerves; medial nerve narrow, percurrent; the lateral thin but distant, arched toward the borders, at a broad angle of divergence from the medial nerve, forking once near the base, the upper branch forking again near the borders, which are undulate by contraction according to the undulations of the surface.

As I have seen only of this fern the specimen figured; its reference to Brongniart's species is uncertain. This last is represented by large parallel secondary pinnæ with close linear-oblong obtuse decurring pinnules, to which the leaflet described above is comparable only by the general facies, the undulations of the surface and of the borders and the nervation. Brongniart compares his species to *Pecopteris punctulata* which is the polymorphous *Callipteris conferta*, Goepp., of which Weiss, "Foss. Fl." has described and figured six different forms as sub-species and some varieties, one of which, var. *sinuata*, of the sub-species *progressa*, p. 80, Pl. VII, Fig. 3, l. c. has for synonym *Pecopteris sinuata*, Brgt. Some of the terminal pinnæ of that variety have a marked likeness to that of Pl. XCIX, Fig. 3, but all of them, in the figures of Weiss as in those of Brongniart, are decurring at base and have the nervation differently marked.

Habitat—The specimen is in a concretion from Mazon creek. Lacoe's, No. 493.

CALLIPTERIDIUM RIGIDUM, *sp. nov. Pl. XCIX, Figs. 1-2a.*

Primary pinnæ large, with broad rachis; secondary pinnæ, at right angles, parallel, rigid, distant, linear, or very gradually narrowed upward, pinnately divided; pinnules close, alternate, oblong, narrowed to an obtuse apex, connate at base one fourth of their length, also at right angles to a thick flat rachis and not decurrent; nervation coarse; medial nerve thick, effaced under the apex; lateral nerves curved in passing to the borders, forking once near the base, the upper branch forking once again near the border as in the preceding species.

The primary rachis is 5 m. m. in diameter; the secondary pinnæ, 25 m. m. distant, are exactly at right angles, on a straight rigid flat rachis, scarcely narrowed upward, at least in the larger pinnæ, which are more than 10 c. m. long.

The pinnules, also comparatively large, 1 c. m. long, half as large (proportionately smaller in the smaller pinnæ), are undulate on the borders and also upon the surface, by the impression of the lateral nerves which are thick and curved back in passing to the borders, their divisions being effaced or merely seen in a few scattered thin leaves. The pinna, Fig. 2, is apparently the upper part of a frond of the same species. The divisions of the veins are distinct, as described above, but the undulations of the borders and surface are scarcely marked.

The affinity of this fern is not distinct. It is evidently a *Callipteridium*, related to *C. orientale*, Schenk, 9th mem. on the plants of the Carb. of China., p. 227, Pl. XXX, Fig. 5—9a., which is a smaller plant with narrower pinnules, but with a nervation of the same character.

Habitat—Olyphant, Penna., Lacoe's, No. 716.

CALLIPTERIDIUM DOURNAISII ?

Pecopteris Dournăisii Brgt., " Veget. Foss." p. 282, Pl. LXXXIX.

The plant which I refer with some doubt to this species is represented by 3 small specimens, all from the same locality and mere fragments. One is that of an ultimate pinna of the middle or upper part of a frond, with lanceolate oblique undulate obtuse pinnules, 1 to 1½ c. m. long, 5–7 m. m. broad; another has two linear pinnæ 12 to 14 m. m. broad, obtusely, deeply and broadly lobate, and a third is the top of a pinna with a short narrow lanceolate obtuse ultimate pinnule. The specimens are too small to show by the forms of the pinnules and pinnæ close relation or identity with the European species, but the nervation is of the same type. The strong medial nerve is flexuous, distinct to the apex ; the lateral nerves are oblique, distant, filiform and very distinct, sometimes forking once only, more generally once near the base and one of the branches or both once again near the borders. In the narrow undulate pinna, the veins obliquely attached to the rachis are at right angles to the base, parallel, distant ; those of the lower side of the medial nerve forking once only at base, those of the upper side forking twice, a nervation much like that of *Odontop-*

teris Schlotheimii, Brgt. As Brongniart says of his species that it is so closely allied to *A. lonchitica*, that it may perhaps be a variety of it, that described above may be a new species as the nervation is far different from that of *A. lonchitica*.

Habitat—Campbell's ledge of Pittston, Lacoe's, No. 995.

CALLIPTERIDIUM GRANDINI.

Pecopteris Grandini, Brgt., "Veget. Foss." p. 286, Pl. XCI, Figs 1–4.

Secondary pinnæ long, close, parallel; pinnules comparatively long, somewhat turned outside, free to near the connate base, contracted below the middle and more enlarged toward the obtuse or blunt apex; medial nerve broad, abruptly vanishing under the apex; lateral veins very close, oblique and arched in passing to the borders, forking once from the middle or from above the base.

The substance of the fern is sub-coriaceous, the surface of the pinnules being inflated or convex. The specimen represents only an ultimate pinna.

Habitat—Buck Mountain, Locust Run, Columbia Co., Penna., No. 1293 of the collection of the National Museum.

ALETHOPTERIS, *St.*

"Coal Flora," p. 175.

ALETHOPTERIS CRASSA.

Sp. nov., Pl. XCVIII, Figs. 5, 5a.

Leaf bi-pinnate; primary and secondary rachis very thick, distinctly regularly lineate-striate lengthwise; pinnæ attached at right angles, but slightly curved upward, distant; pinnules also at right angles, sessile by their whole base, scarcely connate at base, but contiguous to the middle, linear, narrowed to an obtuse apex; medial nerve thick, nearly percurrent; lateral veins at right angles, very thin and close, parallel, scarcely forking, apparently simple.

A fine species known by the figured specimen only. The primary rachis is 14 m. m. broad, flat, distinctly thinly striate, the striæ being parallel, separated by thinner lines, like the veins of leaves of *Cordaites ;* the rachis of the lateral pinnæ 4, m. m. broad, is also flat, less distinctly striate ; the pinnules 2 c. m. long, 5 m. m. broad at base, are linear or nearly so to near the apex where they are gradually narrowed in rounding to an obtuse apex. The medial nerve is broad but not deep or deeply marked ; the veins extremely thin and close, scarcely discernible with the lens, are apparently simple, at right angles and not at all oblique at base, so that the leaflets appear like small pinnules of *Tæniopteris ;* the surface is flat, rugulose, as seen Fig. 5a ; the substance thick, coriaceous.

Habitat—Cannelton, Penna. Lacoe's, No. 1007.

ALETHOPTERIS MAXIMA, *Andrews.*

'*Coal Flora,*" *p. 187.*

Pinna or frond simply pinnate; pinnæ or pinnules close to each other, very oblique and slightly curving upward, parallel, alternate, long, linear-lanceolate, entire, attached by the whole base and decurring by the prolongation of the lower basilar side; rachis comparatively broad, irregularly striate by the exposition of the fibrous tissue by maceration and compression.

This species was established upon a detached ultimate pinna or pinnule. A specimen discovered later represents a large branch bearing pinnules. The main rachis is 7 m. m. broad at the base of the pinna, part of which, 18 c. m. long, is preserved. It bears on each side 8 simple ultimate pinnæ 7–7½ c. m. long, 13–14 m. m. broad near the base, not contiguous on the borders, but parallel, linear-lanceolate, apparently obtuse at the apex (the upper part is nearly effaced by maceration), abruptly decurring to the rachis in joining it and connate to the upper border of the leaflet underneath. The nervation is very close, as described by the author, the lateral nerves forking once or twice and reaching the borders at right angles, not turning upward (as described and figured by the author), but joining the thick medial nerve by a downward curve.

Habitat—Rushville, Ohio, lower carboniferous measures. Lacoe's, No. 314.

PSEUDOPECOPTERIDEÆ.

"Coal Flora," p. 189.

PSEUDOPECOPTERIS, *Lesqx.*

"Coal Flora," p. 190.

The essential group of the genus, that of the *Dicksonioidæ*, has now for its characters the authority of the fructifications, which have been described and finely illustrated in a memoir of Dr. J. F. Sterzel.* He describes the fructification as "*Sori marginal, round, placed at the end of a simple branch of the nerves, near the base of the declining (katadrome) borders of the lobes; the borders of the lobule bearing the sori being reflexed and the parenchyma of the fertile lobule being more or less reduced.*" As recognized by the author, these fructifications have an affinity to those of the *Dicksoniæ* of the present epoch.

PSEUDOPECOPTERIS DIMORPHA, *Lesqx., Pl. XCVIII, Figs. 4, 4a.*

"Coal Flora," p. 201, Pl. XXXV, Figs. 1–6.

The specimens figured in Coal Flora l. c., are all more or less disfigured by the expansion of the pinnæ and pinnules to one side and their contraction to the other; therefore, their true features are not distinctly seen. The fragment figured now, represents a part of the plant in its normal state; the pinnæ linear, the ultimate divisions or pinnules attached at right angles and by the middle to a narrow rachis, oblong-lanceolate in outline, pinnately lobed; lobes 3–4 on each side, broad, obtuse, the terminal also obtuse or half round. That part represents the middle of a long pinna, like that of Pl. XXXV, Fig. 1 of the Coal Flora, or a fragment like Fig. 5 of the same plate, only larger, showing the frond to

* *On Dicksoniites Pluckneti*, "Bot. Centr. Blatt.," XIII, No. 8–9, 1883.

have been a very large one and the species closely related, by the form of its pinnules, to *P. speciosa*, Lesqx., l. c. Pl. LI, the nervation also being of the same character. The relation of this species is also marked with *P. Pluckneti*; and indeed, most of the species of this group, as I have remarked it already, extremely variable as they are, pass by multiple forms from one to the other, rendering their limitation very difficult.

Habitat—The specimen figured is from Olyphant; there is still another like it, from the Orchard vein, mine F, Plymouth, Pa. Lacoe's, No. 349*a* and 349*b*.

PSEUDOPECOPTERIS DECURRENS, *Leqx.*

"Coal Flora," p. 209.

This species was known only from a single specimen, figured and described in "Geol. of Penna.," as quoted in "Coal Flora." A new one, recently procured, represents the species partly as it has been figured, partly with a difference essentially marked in the pinnules which are close, smaller, 2–3 m. m. long, 1½ m. m. broad, with the medial nerve more strongly marked. There are also some fragments with longer pinnules, 5–7 m. m., more or less distant, decurring at base as in the normal form, lanceolate, obtuse, lobate on the borders, with the lowest pinnules pinnately lobate, passing thus to a pinnatifid division, as in some species of *Pecopteris* or *Gleichenia*. For this reason the species should be referred to the first section of the genus, that of the *Gleichenites*.

Habitat. The new specimen has been procured at Alton, Penna. No. 931 of Lacoe's cabinet.

PSEUDOPECOPTERIS NUMMULARIA, *Gutb. Pl. CIII, Figs. 1–3.*

Gutb. "Abdrück," p. 43, Pl. IV, Fig. 5; X, Fig. 7, 8: XI, Fig. 8.

Fronds large, dichotomous; pinnæ large, bipinnatifid; secondary pinnæ open, nearly at right angles, alternate, equally distant, linear-lanceolate, comparatively long; pinnules inclined upward, ovate-lanceolate, sessile by the base of the medial nerve, not decurring, alternately lobate; lobes ovate, three pairs on each side, half round, at right angles to the medial nerve, connate at base, the terminal small, oval or obtusely pointed; medial nerve flexuous, pinnately divided; lateral nerves dichotomous, forking 2–4 times; surface smooth.

The description of this species, made from the specimens we have on hand, does not agree entirely with the figures and descriptions of European authors. In Gutb., l. c., Pl. X, Figs. 7 and 8; XI, Fig. 3, the pinnules are longer, 4–5 lobate on each side, the lobes round, narrowed, subpedicellate and generally decurring at base, and the veins radiating and dichotomous from the base. In Heer, "Fl. fos. Helvet," Pl. XIV, Figs. 6–6*b*., where the species is also figured and described, the form of the pinnules and of the lobes is the same as upon the American specimens; but the lobes are more distinct, separated to the base, where they are somewhat narrowed and decurring. The nervation is as figured in Pl. CIII, Fig. 1*a*. The only representation of a species agreeing in its forms with those of the American specimens is that of Roehl "foss. Fl. Westph.," Pl. XVI, Figs. 6 and 6*a*., a large tripinnatifid pinna, which he refers to *Sphenopteris stipulata*, Gutb. But this last species, according to Geinitz, is far different from *S. nummularia* being for him identical with *S. irregularis*. Compared to *S. stipulata* of Gutb., the American forms present still more marked differences and therefore it is questionable if it should not be described as a new species. We may even have two different species in those specimens. For all those from the subconglomerate measures preserve as their essential characters the form of the lobes half round, connate at base, not decurring, though extremely variable in the size of the pinnules, as seen Figs. 1–3 of Pl. CIII. But in some specimens from Cannelton and in those from Clinton, Mo., the pinnules are somewhat larger, the lobes nearly round, 3 rarely 4 on each side, distinct to near the base, and somewhat decurring.

Habitat—1st type. Subconglomerate measures. Campbell's ledge, Pittston, R. D. Lacoe; Arkansas, Harvey. 2nd type. Coal above conglomerate, Cannelton, Penna., I. F. Mansfield; Clinton, Mo., Dr. J. Britts.

PSEUDOPECOPTERIS OBTUSILOBA.

Sphenopteris obtusiloba, Brgt., "Veget. Foss.," p. 204, Pl. LIII, Fig. 2.

Leaves bi or polypinnate; rachis subflexuous; pinnæ long, linear-lanceolate, rapidly narrowed upward, triangular in outline; pinnules short and enlarged sidewise, often reniform or sub-triangular, obtuse or truncate at the apex, the lower deeply trilobate, the upper simple; lobes sub-equal, round-ovate, sometimes trilobate; nerves distinct, dichotomous, a number of them passing into each lobe.

Though I have not seen this species in the former years of my explorations it appears to be locally abundant especially in the sub-conglomerate coal measures. It is extremely variable in the subdivisions of the pinnules which sometimes are all more or less distinctly trilobate in each pinna, but sometimes also entire in the upper part of the pinnæ, gradually smaller upward and most variable in forms. One of the more marked varieties, found sometimes with the normal form, is that which representing *Sphenopteris dilatata*, Ll. and Hutt., "Foss. Fl.," Pl. XLVII, has the pinnules either simple, entire, or three-lobed, the lobes being generally enlarged at the apex and the pinnules, when simple, widely apart and contracted at base to a short petiole. The nervation is that of the normal form of *P. obtusiloba*, the primary nerve forking at and above the base in two or three branches, each entering one of the lobes, and the branches forking again once or twice within the lobes according to their size. The rachis, in the variety as in the normal form, is smooth, somewhat flexuous; the secondary pinnæ very long, as long as 13 c. m., rigid, parallel, at right angles, lanceolate, the tertiary inferior pinnæ (or pinnules) at least 2 c. m. long, gradually shorter upward, the lowest, with 4 rows of alternate lobes, the lobes 3–4 lobate, or in the upper pinnules bi-lobate or entire, the terminal small, obtuse, often crenate. In the normal form, the secondary pinnæ are short, only 4–5 c. m. long, the tertiary ones are five lobate, the lower lobes bi- or tribolate, the upper entire. But both forms of pinnules with the same sub-divisions are sometimes observed upon the same specimen.

Habitat—Specimens of the normal form are from W. Va. and from Clinton, Mo. Lacoe's, No. 348. Those of the variety from Wyoming mines, W. Va. Lacoe, No. 261.

PSEUDOPECOPTERIS ANDRAEANA.

Sphenopteris Andraeana, Roehl, "Foss. Fl. Westph.," p. 62, Pl. XXII, Fig. 6.

Lower pinnules of the ultimate pinnæ 2–5 lobed, the upper simple; lobes alternate, obovate, oblique, two or three on each side; upper pinnules gradually shorter with one lobe on each side, the terminal larger, all denticulate on the borders; nervation dichotomous-flabellate from the base of the primary nerves, immersed into the epidermis, obsolete.

I have only seen small fragments of this peculiar species which, in its aspect, the general outline of the pinnæ, the form of the pinnules and of the lobes, even in the nervation, has the greatest affinity to *P. obtusiloba.* The essential difference is in the thick substance of the pinnules wherein the veins are imbedded and therefore often obsolete, not prominent, and in the distinctly denticulate borders. The main rachis is flat on the borders, channeled in the middle.

Habitat—Blacksburg and Malden mine, W. Va. Lacoe's, No. 817.

PSEUDOPECOPTERIS MACILENTA, *Ll. and Hutt.*, *Pl. XCVIII, Fig. 2.*

"Coal Flora," p. 219.

I have figured here the specimen described "Coal Flora," l. c., representing one of the forms of the species which, polymorphous like the other ferns of this genus has characters somewhat different in the divers parts of the same plant. With the fragment figured, the description will be more easily understood.

Habitat—The specimen figured is from Cannelton, Pa., procured by Mr. I. F. Mansfield.

PSEUDOPECOPTERIS HISPIDA, *Sp. nov.*, *Pl. XCVIII, Fig. 3.*

Leaf pinnately divided; secondary pinnæ long, linear-lanceolate, narrow; tertiary pinnæ (pinnules) distant, sessile by the abruptly narrowed or obliquely truncate base, oblong-lanceolate, obtuse, pinnately lobed; lobes short, obtuse, the terminal broader and obtuse; surface hispid or scaly; nervation obsolete.

Of this species I have seen only the specimen figured. It is a straight, rigid pinna, with a narrow rachis and alternate distant compound pinnules, 9 on each side, inclined upward, the lower 18 m. m., the upper 8 m. m. long, 1 c m. or less broad at base, regularly pinnately divided to below the middle of the lamina into short obtuse lobes, turned upward, all very entire. The surface is punctulate or rough, as by scars of scales or hairs destroyed by maceration. The species has a distant relation to the preceding, but is evidently different.

Habitat—Cannelton, Penn'a.

PECOPTERIDEÆ.

"Coal Flora," p. 221.

PECOPTERIS, *Brgt.*

"Coal Flora," p. 223.

PECOPTERIS ROBUSTA, *Lesqx.*, *Pl. XCIX, Figs. 4, 4a.*

"Coal Flora," p. 229, Pl. XXXIX, Figs. 7-8.

The small specimen now figured represents the species in a somewhat different character, especially in the form of the leaflets which are not all connate at base, or attached to the rachis by their whole base, but some of them rounded to the point of attachment by the middle of the pinnules and more or less distant. That the difference is not of specific value is seen in the peculiar fact that on the left side of the rachis, the pinnules are decurrent, connate at base, or attached by their whole width, while on the right side they are rounded at base, joined to the rachis by the middle and disconnected. The form of the terminal small oval pinnule and the nervation are the same in both specimens.

Habitat—The specimen figured is from Mazon creek, in concretions. The fragments described l. c. p. 229, were obtained at Port Griffith, near Pittston, from a bed of shale considered as above coal F.

PECOPTERIS ASPIDIOIDES, *Brgt.*

"Hist. d. Veget. Foss.," p. 311, Pl. CXII, Fig. 2. P. arborescens, var. aspidioides, Lesqx., Coal Flora, p. 231.

This species has been considered by Schimper and other European authors as a variety of *P. arborescens.* I have formerly admitted the same opinion; but from numerous specimens constantly representing the same character in all the localities where they have been found and where the true *P. arborescens* has not been observed, I now admit the species as distinct. It differs from *P. arborescens* by its generally larger size, the pinnæ broader, the pinnules longer, the terminal one very small, nearly round or bi-trilobate, the lateral veins distinctly curved backward and the strongly striate rachis.

Habitat—Numerous specimens have been received from Ottawa, Kansas; from Cannelton, Penna., and also from Mazon creek, all localities of the lower coal strata above the conglomerate, while *P. arborescens,* very abundant at the horizon of the Pittsburgh coal and above, has been very rarely found at a lower stage.

PECOPTERIS QUADRATIFOLIA, *Lesqx., Pl. C, Figs. 1, 2.*

"Coal Flora," p. 234.

This species appears to be variable, like *P. arborescens,* which it resembles. The pinnules are sometimes very small, 1–2 m. m., but sometimes as long as 5 m. m., always abruptly truncate at the apex or nearly square, very close to each other, contiguous on the borders, though not connate. The rachis is broad, distinctly punctulate, the nervation generally obsolete.

Habitat—Mostly found, like the preceding, in the lower coal strata above the conglomerate. Colchester, Morris, Connelton, etc.

PECOPTERIS SERRULATA, *Hartt. Pl. CI, Figs. 4–8a.*

Frond tripinnately divided; pinnæ oblique, alternate, distant, lanceolate, variable in length; pinnules ovate-lanceolate, more or less deeply dentate or serrate; lateral nerves oblique, simple or forking at the apex; fructifications in round very small sori, placed in single rows quite near the borders of the pinnules.

The definition of this species as given by Schimper "Pal. veget.," III, p. 495, and that of Dawson in "Geol. Survey of Canada," p. 55, are both incomplete, as made from too small specimens. The frond of this species has the same kind of divisions as that of the *Cyatheites*, the ultimate pinnæ being either simply or doubly pinnate, according to their position upon the fronds. In Fig. 7, the lobes of the pinnæ are entire ovate obtuse small pinnules, 2–3 m. m. long, separated to their base. In Figs. 5 and 6, these lobes are 3–5 m. m. long, merely serrulate, shorter or longer, enlarged and connate at base. In Fig. 4, the lobes, like small ultimate pinnæ, are 10 to 15 m. m. long, lanceolate, deeply dentate, with the medial nerves pinnately divided and the veins forking at the apex. In all the numerous specimens which I have seen of this species, the veins are sometimes simple, sometimes forking once or twice, as in *P. plumosa* or *dentata*, Brgt. The fructifications are placed quite near the margins of the leaflets, in small round sori, generally covered, at least in the specimens I have seen. As far as can be seen, the only marked difference between this species and some of the varieties of *P. dentata*, is the smallness of all its parts and the slenderness and flexibility of the rachis.

Habitat—Sub-conglomerate coal measures, abundant—Arkansas, F. L. Harvey, also at Tracy, Tenn., Dade, Georgia.—Lacoe's, No. 859.

PECOPTERIS OBSOLETA, *Sp. nov. Pl. CXI. Figs. 29, 29b.*

Leaves bi-pinnately divided; ultimate pinnæ open, nearly at right angles, alternate, somewhat distant, parallel, linear-lanceolate, narrow; pinnules small, oblong, very obtuse, contiguous to near the apex, connate at base, smooth, without trace of a medial nerve; fructifications in round or transversely oblong sori, in two rows, quite near the borders; upper pinnæ entire, obtuse, somewhat distant, rounded at base to the rachis.

It will be seen in the figure that as it is often the case in species of *Pecopteris*, the pinnules of the ultimate pinnæ become gradually connate, simple, the pinnæ becoming large pinnules. In this species, the upper leaflets are not connate but rounded at base and oblong-obtuse, somewhat contracted in the middle, like those of *P. Strongii.* The texture of the leaves of this species is extremely thin and delicate, as most of the branches rachis and leaflets are totally destroyed by maceration, leaving upon the specimens the sori only, whose position indicates the outlines of the pinnæ and pinnules. By their peculiar distribution and the form of the sori, the species resembles what Weiss has published as *Ptychocarpus hexastychus*, "Foss. Fl.," p. 95, Pl. XI, Fig. 2.

Habitat—Cannelton, Penna.; I. F. Mansfield.

PECOPTERIS CARRII, *Sp. nov., Pl. C, Figs. 3, 3a.*

Pinnæ rigid, bipinnatifid: lateral pinnæ nearly linear, at right angles to the narrow straight rachis; pinnules alternate, distant, oblong-lanceolate, abruptly pointed, decurring by the prolonged base bordering the rachis, distantly 3-4 dentate on each side, pinnately nerved; primary nerves thin but very distinct; the secondary alternate, oblique and curving in passing to the teeth, the lower pair only forking once above the middle, the other simple.

I do not know any species to which this could be compared. The fragment of a pinna which represents it is rigid and regular in all its divisions. The alternate pinnæ, at right angles, are 8 m. m. distant, parallel, the lower 3 c. m. long, the upper 2 c. m.; but they all appear broken and were probably much longer; the pinnules inclined outside, all equal and equidistant, 4 m. m. long, 2 m. m. broad at base, are decurring along the rachis, which becomes narrowly winged by the basilar prolongation of the leaflets. The borders of the pinnules are cut into distant short teeth turned outside, separated by shallow sinuses, and the lateral nerves alternately curve in passing obliquely from the mid-rib, each division corresponding to one tooth and entering it, as seen Fig. 3 *a* enlarged.

Habitat—Mazon creek; in nodules; only one specimen. Communicated by J. C. Carr of Morris

PECOPTERIS GEORGIANA, *sp. nov. Pl. XCVIII, Fig. 6, 6a.*

Ultimate pinnæ at right angles, linear; rachis thick, inflated, bordered by a narrow lamina cut into very small pinnules connate to the middle, regularly turned outside like the teeth of a saw, bidentate at apex; primary nerves simple at base, inclined, forking at the apex, each branch entering one of the teeth.

This species, abundantly, found in the sub-conglomerate coal measures, is closely allied to *P. erosa*, "Coal Flora," p. 255, Pl. XLIV, Fig. 1-3, but appears different by its thick coriaceous substance inflated and like cartilaginous along the broad rachis, and by the lobes more inclined outside, separated to the middle and only bidentate, the teeth being short and sharply pointed. In *P. erosa*, the pinnæ are short, 2 to 3½ c. m. long. In this form or species, they are much longer; for other specimens obtained later, have fragments of leaves at least twice as long as those figured. It is, however, remarkable that in the same localities or in the sub-conglomerate measures, a form of this species or an allied species has been found with the divisions or lobes acutely 4 to 5 dentate, the medial nerves pinnately forking in two branches which, either simple or forking, enter each one of the teeth. This form evidently corresponds to that of *P. cristata*, which, as seen in the description of the species, "Coal Flora," p. 256, seems to be a variety of *P. erosa*.

The group of Pecopteris (Crestate) has been separated by Zeiller into a new genus, *Grand' Eurya*, apparently identical to the genus *Saccopteris* established later by Stur.

Habitat—Dade, Georgia, also found in Tennessee. Lacoe's, No. 855, 855*a*.

PECOPTERIS ORNATA, *Sp. nov., Pl. CXI, Figs. 30–30b.*

Fragment of a bipinnatifid lanceolate leaf; ultimate pinnæ slightly oblique, rigid, short, narrowly lanceolate, contracted at their point of attachment to the main round rachis, slightly decurring, pinnately cut along the borders into short obtuse lobes filled by 2 to 4 round open sori, exposing to view the small globular sporanges disposed in a circle around a central point, Figs. 30, 30*b*, as in species of *Hemitelea* of our time. The secondary rachis is round, smooth, and very thin, the lateral veins, obliquely diverging from it, enter each of the lobes and are apparently simple. The species is in close relation to and of the same group as *P. stellata*, Lesqx., "Coal Flora," p. 260, Pl. XLVIII, Figs. 7, 7*b*—and *P. lyratifolia*, ibid., p. 259, Pl. XLVIII, Figs. 4–5*b*.

Habitat—Mazon Creek, in nodules No. 1008 of Lacoe's cabinet.

SOROCLADUS, *Lesqx*

"Coal Flora," p. 327.

SOROCLADUS SAGITTATUS, *Lesqx., Pl. C, Figs. 4, 5.*

"Coal Flora," p. 329, Pl. XLVIII, Figs. 10—10b.

The specimens figured show more evidently the peculiar characters of that fruiting frond, first in the mode of division of the rachis which is bipinnate; then in the mode of attachment of the capsules containing sporanges, which are sessile or short pedicellate, and then, Fig. 5, in the evidence of the nature of these organs as the fructification of a fern. It is regrettable that the sterile part, top of Fig. 5, is not larger, as its characters, the form and the nervation of the pinnules, can not be satisfactorily recognized. The characters appear to be those of the fern described as *Pecopteris abbreviata*? Brgt. "Coal Flora," p. 248, Pl. XLVI, Figs. 4–6; but it is probably a new and distinct species, as the specimens of the so-called *P. abbreviata* sent to Europe for comparison, have not been recognized as referable to Brongniart's species.

Habitat—Found only in the concretions of Mazon Creek.

PHENOPTERIS, *Brgt.*

"Coal Flora," p. 268.

SPHENOPTERIS (PECOPTERID).

"Coal Flora," p. 269.

SPHENOPTERIS CRISTATA, *St. Pl. CII, Figs. 1, 1a.*

"Coal Flora," p. 273.

Pecopteris cristata, Brgt., "Veget. foss.," p. 356, Pl. CXXV, Figs. 4, 5.

The specimen figured agrees with the description in Coal Flora, l. c. It differs slightly from the European form by the lobes a little more deeply cut and the teeth more acute. The pinnules are also a little longer. As the specimens have been somewhat distended, as they generally are in the Coal Measures of Rhode Island, the slight differences are unimportant. Other specimens have shorter lobes and shorter pinnules.

Habitat—Rhode Island, not rare. Collection of Lacoe, No. 828.

SPHENOPTERIS CRISTATA, *var. minor*, *Lesq.*, *Pl. CIV, Fig. 5, 5a.*

Leaves tripinnatifid; pinnæ long, linear-lanceolate; pinnules ovate-oblong, abruptly narrowed at base, pinnately lobate; lobes trifid, sharply dentate; nerves pinnate, the divisions trifid, entering the teeth.

This form differs from the preceding not only by the smaller size of all its parts, but essentially by the trifid divisions of the lobes of the pinnules, each of the divisions being acutely bi- or tri-dentate. The secondary rachis is strong and rigid, the pinnules only half as long as in the normal form. It may be a distinct species.

Habitat—Cannelton, No. 897, of Lacoe collection.

SPHENOPTERIS COMMUNIS, *Sp. nov., Pl. CIV, Fig. 1, 1a.*

Fronds deltoid-lanceolate, tripinnatifid; pinnæ at right angles on a narrow rigid winged rachis, narrowly oblong-lanceolate, alternate, parallel; pinnules slightly inclined upward, decurring to the rachis by a narrow border, oblong-lanceolate, irregularly pinnately lobed; lobes oblong, obtuse, cut to below the middle or nearly connate in the upper pinnules, entire or indistinctly crenulate on the borders; medial nerves somewhat flexuous, pinnately divided; veins rarely simple, generally forking from the middle, more or less incurved along the borders, sharply marked.

This species is so much like *S. pseudo-Murrayana*, described "Coal Flora," p. 271, that I was at first inclined to consider it as a variety; the more so, as Brongniart's species, *Pecopteris Murrayana*, is represented under divers forms. There is, however, a marked difference in the rigidity of the pinnæ, in the nearly coriaceous texture of the pinnules and in the nervation, the veins being all distinctly marked, rather prominent and the veinlets generally forking. The pinnules all have the same shape and the same kind of division, the lowest, 9 to 13 m. m., according to their position, have the lobes separated to below the middle, decurring, entire or obtusely crenulate, the lobes of the upper ones are connate nearly to the apex, and therefore the upper pinnules appear as regularly crenate or undulate on the borders. The divisions of the stipe, both primary and secondary, are narrow and bordered by a narrow band formed by the decurring base of the pinnules. The nervation is distinctly seen upon the enlarged Fig. 1*a*.

Habitat—Very common in the sub-conglomerate coal of Arkansas, Male's coal, also sent from Dade, Georgia, and from Tracy, Tennessee, No. 442, Lacoe's collection.

SPHENOPTERIS PLICATA, *Lesqx., Pl. CXI, Fig. 28.*

Lesqx. "Geol. of Penna., **1858,** *p. 862, Pl. IX, Fig. 3; "Coal Flora," p. 292.*

Fragment of a long linear-lanceolate pinna with alternate oblique pinnules decurring and connate at base, oblong, obtuse in outline, pinnately irregularly lobed; lobes short, merely marginal, round or crenulate, the terminal half round, small; primary nerves decurring, thin but distinct, ascending to the apex, pinnately branching; branches forking once or twice, thin.

The species appears to be very rare. I found the fragment published in Geol. of Penna., l. c. at an old abandoned shaft behind New Philadelphia. The one figured here is from a nodule of Mazon creek, No. 997 of Mr. Lacoe's collection.

SPHENOPTERIS (PROPER.)

"Coal Flora." p. 274.

SPHENOPTERIS GRAVENHORSTII, var. B., *Brgt., Pl. CI, Figs. 1—1b.*

Brgt. "Vegt. Foss." p. 191, Pl. LV, Fig. 3.

Primary pinnæ large, lanceolate in outline, with a flexuous canaliculate rachis; secondary pinnæ alternate, distant, oblique at base, recurved in the upper part; tertiary pinnæ open, nearly at right angles, lanceolate, alternately pinnately lobed; lobes 5–7 on each side, oblique, oblong-lanceolate, decurring, connate at the base only, bi- tri-obtusely crenate; medial nerve flexuous, pinnately branched, each branch entering one of the lobes and forking once or twice, the veinlets passing up to the points of the teeth.

There are two specimens of the species; the smaller of the two is figured. The characters of both are the same. The rachis is canaliculate in the middle or narrowly flat-margined, distinctly flexuous, comparatively narrow, about 2 m. m., that of the secondary pinnæ being nearly as broad. The lobes of the pinnules are short and obtusely dentate or crenate; but the lobes are distinct to the base. In the normal form, figured by Brgt. l. c., the teeth of the lobes are more distinctly marked and more acutely dentate. The author remarks that var. B. of this species has the pinnæ shorter, more deeply pinnate, the rachis sinuous and the pinnules deeply partite, with more acute lobes. I refer the American form to this variety with which it agrees by the flexuous rachis, but the pinnules are not more deeply divided nor the lobes more acute than in the normal form.

Habitat—Dade county, Georgia; sub-conglomerate coal; found only in two specimens. No. 262 of Lacoe's cabinet.

SPHENOPTERIS HYMENOPHYLLOIDES, *Brgt. Pl. CII, Fig. 2.*

Brgt., "Veget. foss.," p. 189, Pl. LVI, Fig. 4.

Ultimate pinnæ linear-lanceolate; pinnules open, less oblique than in the last species, decurring to the narrow winged rachis, ovate-lanceolate, pinnately lobate above the constricted base; lobes bi- tri-acutely dentate, the terminal teeth entire, sharply acute.

This form differs from the European by the broader wing of the rachis. The specimen, much smaller than that of Brongniart, apparently represents the upper part of a frond or of a secondary pinna with ultimate pinnæ preserved upon one side only. The main rachis is narrow, the pinnæ about 5 c. m. long, the pinnules a little narrowed near the base, ovate above, average 1 c. m. long, 5 m. m. broad, all sharply bi tri dentate, connate above the base; lateral veins bifid near the apex, their divisions entering the teeth. The venation is of the same character as that marked for the preceding species in Fig. 1*a* enlarged.

Habitat—Cannelton, Penn'a.; I. F. Mansfield.

SPHENOPTERIS BRITTSII, *Lesqx., Pl. CII, Figs. 3–4a.*

"Coal Flora," p. 277, Pl. LV, Figs. 2–2b.

The specimen figured and described in the flora, too small and imperfect, does not sufficiently show the characters of this fine species.

The stipe is large, a fragment of it, apparently at the base of a frond is 3 c. m. in width, flattened. The rachis in all its divisions is smooth (not punctate as described from fragments of stems probably pertaining to *S. mixta*). The primary pinnæ are very large, flexuous, tripinnatifid, with a rachis 2 m. m. broad; the secondary are alternate, 3 c. m. distant, flexuous, linear-lanceolate, the lower 12 c. m. long or more, with 20 to 24 pairs of alternate oblique pinnules; tertiary pinnæ, 3 c. m. long at the base, gradually shorter toward the apex, all linear-lanceolate. These are composed of 6–12 pairs of alternate oblique oblong lobes, quite close, contiguous near the base, somewhat enlarged upward, crenate or obtusely dentate at the apex, rugose or verrucose on the surface, as seen Figs. 3*a* and 3*b*. The veins of the pinnules are pinnately forked, the veinlets in the lobes dichotomous, the divisions entering the teeth. The fructifications, Figs. 4 and 4*a*, are round inflated or flat sori, at the end of the upper branches of the veins, near the borders, and quite distinct from the small tubercles covering the leaves; the sporanges, however, are not visible. The fructifications resemble those of species of *Dicksonia*.

Habitat—Large specimens have been procured near Nelsonville, Ohio, from coal No. 6, by the kindness of Superintendent Roy, for Mr. Lacoe's collection, No. 372. Fruiting specimens have also been sent from Clinton, Missouri, by Dr. Britts.

SPHENOPTERIS INÆQUILATERALIS, *Sp. nov., Pl. CIII, Figs. 4—5a.*

Pinnæ bipinnatifid, apparently large, linear-lanceolate; secondary divisions long and narrow, linear, distant, open or at right angles to the rachis, flexuous; pinnules alternate, distant, short, decurring on the narrowly winged rachis, enlarged at base by the expansion of the lower lobes, lanceolate, blunt at apex, trilobate or tridentate on both sides; lower lobes enlarged on the upper side, denticulate on the borders, the others short, connate to near the apex, being merely obtuse teeth; nerves slightly decurring at base alternately forking; veinlets all prominent and thick, the lower of the upper side trifid, the others simple.

This species, which I cannot compare to any other, is remarkable for the unequilateral shape of its small pinnules which are nearly triangular, and for the strong very distinct nervation. The main rachis is comparatively thick, 3 m. m., the secondary very narrow, flexible, scarcely 1. m. m.

Habitat—Sub-conglomerate coal of Arkansas, F. L. Harvey.

SPHENOPTERIS (HYMENOPHYLLITES.)

"*Coal Flora*," *p. 281.*

SBHENOPTERIS HARVEYI, *Sp. nov.*, *Pl. CIII, Figs. 7—7b.*

Fragments of a tripinnatifid frond. Primary rachis comparatively thick, half cylindrical in the middle, flattened on the borders; primary pinnæ open, flexuous, distant, irregularly scattered upon the stone; secondary very oblique, decurring, alternate, erect or curved backward; pinnules or tertiary branches close, decurring, cut to near the base in alternate filaments dichotomously divided into very slender filiform branchlets, either naked or bearing at the apex globose punctiform sori apparently naked, sometimes irregularly grouped.

This fine species is allied to *S. Bronni*, "Gutb. Abdr," p. 37, Pl. V, Figs. 1, 2, described and figured in fruit in "Gein., Verst.," p. 17, Pl. XXIII, Figs. 15 and 16. It differs essentially by the ramification of the segments of the pinnules which are alternate, close, divided to near the base into very thin filiform dichotomous branches, longer and narrower than in the European plants, all of the same thickness, either naked or bearing globose sori at their apex.

Habitat—Sub-conglomerate coal of Arkansas, F. L. Harvey, very rare, No. 370 of Lacoe's collection.

SPHENOPTERIS (DIPLOTHMEMA) TRACYANA, *Sp. nov.*, *Pl. CI, Fig. 2.*

Primary pinnæ lanceolate, with a strong rigid rachis traversed by the flexuous narrow pith or axis; secondary pinnæ long, linear-lanceolate, open or slightly oblique, with a rigid strong rachis; pinnules alternate, somewhat distant and oblique, bi-tri-pinnately divided into lobes cut into 2 to 4 short-linear narrow alternate segments.

A beautiful fern seen only in the fragment figured. The rachis is broad and rigid, but appears to have been composed of a thick bark of soft texture and a hard narrow axis which is alternately inclined to the point of attachtachment of each pinna. The secondary pinnæ, alternate and parallel, 10 to 15 m. m. distant, are rigid, open; the pinnules alternate, 5–7 m. m. distant, oblique, are composed of a rachis thick at base, gradually narrowed upward, and of alternate lobes cut in very narrow generally alternate 2–4 linear obtuse segments, 1 to 1½ m. m. long, less than ½ m. m. broad.

The species is allied to *Diplothmema Schutzei*, Stur, "Culm Fl.," p. 128, Pl. XIII, Fig. 4, differing especially by the smaller size of all its parts, the mode of branching, the secondary pinnæ alternate open, the pinnules and their divisions shorter and more slender. In the European species, the rachis is thick and transversely rugose. Both species, however, are of the same type.

Habitat—Tracy, Tennessee, No. 930 of Lacoe's cabinet.

SPHENOPTERIS DIVARICATA, *Goepp., Sp. Pl. CIV, Fig. 6.*

Cheilanthes divaricatus, Goepp., "Syst.," p. 238, Pl. XII, Figs. 1, 2.
Calymnotheca divaricata, Stur, "Culm. Fl.," p. 165, Pl. XIII, Figs. 1-3.
Sphenopteris elegans, St. not Brgt.

Frond pinnately divided; ultimate pinnæ linear-lanceolate, short, open; pinnules alternate, ovate-oblong in outline, alternately pinnately lobate; lobes short, linear, or enlarged above, truncate at the apex; nervation obsolete.

Prof. Stur who has figured l. c., a splendid pinna of this species describes the nervation as simple or bifid. In the specimens I have seen, no trace of nerves is visible and none are seen on Goeppert's figure. The species is closely related to *S. elegans*, Brgt.; the appearance is however different. The pinnules are more distinctly pinnately lobed, longer and narrower, the lower lobes bifid, the upper linear, narrower, longer, truncate or bifid.

Habitat—Sub-conglomerate coal measures, Dade, Georgia. No. 923, Lacoe's collection.

SPHENOPTERIS ROYI, *Sp. nov., Pl. CIV, Figs. 7–10.*

Rachis broad, flexuous, margined; ultimate pinnæ alternate, short, nearly at right angles; pinnules alternate, oblique, decurring on the winged rachis, oblong or obovate in outline, cut into four to six linear-lanceolate acuminate segments alternate or opposite simple or bifid.

This fine species is known only by small fragments of pinnæ with ultimate divisions at right angles. The rachis is comparatively broad-winged, canaliculate, and flexuous or sub-geniculate. The ultimate pinnæ are short, only 2–3 c. m., mostly broken, one only being preserved nearly entire, 2 c. m. long. The pinnules are close, alternate, inclined outside, from 7 to 14 m. m. long, entire at the decurring base, palmately divided above into 2 to 6 or more linear acute laciniæ, with a thin nerve branching according to the number of the divisions, each entered by one veinlet. In the fragments Figs. 7 and 8, the segments of the pinnules are short, 3 m. m., more obtuse, truncate, apparently broken; in Figs. 9 and 10, they are much longer, bi- or trifid at base, each division being split again to the middle into linear-lanceolate sharply acuminate segments 5 m. m. long or even multifid from above the base and cut into 4 to 6 double laciniæ. The fragment Fig. 10 would seem to be referable to a different species; but there is a gradual elongation of the pinnules and of their divisions upon the different specimens, as between Figs. 8 and 9, and even upon the same specimen, Fig. 10, from the base upward and therefore it is impossible to indicate a character which could authorize the separation, most of the fragments being disseminated upon the same pieces of shale.

Habitat—Rockwood mines, Tennessee. Specimens procured by Mr. Roy, superintendent of the coal mines of Ohio. Lacoe's collection, No. 1037.

SPHENOPTERIS SOLIDA, *Sp. nov., Pl. CI, Fig. 3.*

A fragment of a lanceolate bipinnate leaf. Primary and secondary rachis thick, round, solid, smooth; lateral pinnæ oblique, parallel, rigid, and distant; leaves alternate, either simple, very oblique, decurring, cuneiform, enlarged and very obtuse at the apex, or open, nearly at right angles, much larger, palmately 3–4 lobate, the lobes also cuneiform, obtuse or truncate at the apex; nerves thick, distant, flabellate and dichotomous, diverging from the base.

As far as seen by the specimen, the lower leaves are generally palmately lobed and open, sometimes at right angles upon the main rachis, the others upon the secondary stems are more generally entire, of the same form as the lobes of the lower leaves, cuneiform, obtuse, but some of them also bi- or tri-lobate, the lobes being sometimes enlarged and obtusely lobate or truncate emarginate at the apex. The nerves are distinct, rather inflated, all joined at base, mostly forking once or dichotomous and flabellate.

The relation of this species is not distinctly marked with any other. It may be compared to *S. palmata*, Schp. Pal. Veg. 1, p. 388, Pl. XXVIII, Figs. 1, 1*b*, by the thick rachis and the mode of division of its lower leaves; but that is the only point of affinity; for Schimper's species has the rachis broadly canaliculate and narrowly alate, all the leaves palmately divided and the nervation totally indistinct. It has also the same points of affinity and about the same degree of difference with *Sphenopteris spinosa.* "Geopp. Gatt.," 3 and 4, p. 70, Pl. XII.

Habitat—Mazon Creek, communicated by Mr. Frank T. Bliss.

SPHENOPTERIS OBOVATA, *Ll. and Hutt., Pl. CIII, Fig. 6.*

Ll. and Hutt., "Foss. Fl.," II, p. 75, Pl. 109. Adiantites microphyllus, Goepp. "Syst.," p. 228.

Frond tri-pinnate; divisions alternate, distant; rachis flexuous in all its divisions, sulcate or flat by compression; pinnules very oblique, small, obovate, either distant or close, separated to the base or confluent in their whole length.

In the general appearance, the shape, and the relative and varied position of the leaflets, the American form is similar to that described from English specimens. The only difference may be in the nervation, which, in the specimens I have examined, is generally obsolete. The English authors figure the pinnules (as seen in the magnified fragment) with a thick medial nerve and thin very oblique veinlets derived from the nerve, dichotomous and curving to the borders, as in a *Neuropteris*. The description however says: "No mid-rib can be found of these pinnules, nor any kind of veins beyond a number of parallel striæ which fork occasionally". Upon some detached leaflets, I have observed obscure veinlets, apparently very close, diverging from the base, therefore appearing as described by Ll. and Hutt., but no trace of medial nerve.

Habitat—Sub-conglomerate coal measures of Arkansas; F. L. Harvey.

EREMOPTERIS, *Schp.*

"Coal Flora," p. 292.

EREMOPTERIS CHEATHAMI, *Sp. nov., Pl. CIV, Figs. 2-4.*

Bi-pinnate; rachis round, solid, inflated at the divisions and somewhat flexuous; primary pinnæ lanceolate, the lateral oblique or at right angles, alternate, oblong or linear-lanceolate, obtuse; pinnules alternate or sub-opposite, cuneiform, decurring at base, rarely entire and obovate, generally trilobate or quadrilobate, the lobes connate to above the middle, oblong or obovate, obtuse, entire; veins flabellate, dichotomous from the base of the pinnules, very thin but distinct.

The lower lateral pinnæ about 3 c. m. long, are rapidly shorter upward, passing near the top to trilobate pinnules. In Fig. 4, the lobes appear to be crenulate at the upper border, as in *E. crenulata*, "Coal Flora," Pl. LIII, Fig. 1; but that is a mere appearance caused by the breaking of the slate or by the closeness of narrow lobes. The texture of the leaves is subcoriaceous. In some parts the rachis is transversely rugose. The species is related to *Sphenopteris nervosa*, Brgt., "Veget. Foss.," p. 174, Pl. 56, Fig. 2, a species which has the nervation much stronger. It is a true *Eremopteris*.

Habitat—Sub-conglomerate coal measures, Tracy, Tennessee. No. 1038 of Lacoe's collection.

ARCHÆOPTERIS, *Dawson*.

"Coal Flora," p. 299.

I have had lately the opportunity of examining, for determination of the species, a number of finely preserved large specimens of *Archæopteris*, all obtained from the Red Shale of the Pocono at Meshoppen and along the Susquehanna river above Pittston, Penna. Though the species are not represented in figures in this volume,* I give here an abridged account of the result of my researches. They may facilitate the future study of the beautiful plants of this genus.

The specific characters of *Archæopteris* may be derived:

1st. From the rachis, which is either smooth or striate lengthwise, with or without points or asperity on the surface, or distinctly rough, or transversely rugose.

The transverse rugosity of the stems is however not always distinct. Some stems more or less evidently striate lengthwise are marked here and there by more or less distinct points or small protuberances, especially upon the bark, and these are sometimes so close that it is difficult to ascertain if the rachis is merely striate or if it is rough or transversely rugose.

2d. From the nervation, either thin, obsolete, or distinct. The nervation is generally more distinct near the base of the pinnules or along their upper border.—Taken altogether the nervation has nearly the same character in all the species, being either totally or partly obsolete, or

---

*The specimens which represent new species have been figured, but the plates could not be ready in time. They are the property of Mr. R. D. Lacoe as well as the specimens which he has procured himself.

more or less distinct, even upon the same specimen, and therefore the nervation by itself alone does not afford a definite or reliable character.

3d. From the form and position of the pinnules and from the contexture of their borders, either entire or denticulate, or coarsely dentate, or split.—The leaflets are more or less enlarged upward, exactly wedge-shaped, or more or less broadly obovate. They are entire at the upper borders, or split and laciniate, or denticulate by the projection of the nerves, or more coarsely dentate all around. But, according to the degree of maceration or to the mode of preservation, the borders of the leaves are more or less distinctly crenate or split, and the excurrent nerves are more or less projecting out of the curved outline of the borders.

4th. From the distribution of the flowers, either in the middle of the tertiary branches, placed in series of short racemes taking the place of leaflets which occupy the lower and the upper part of the pinnæ, or occupying by themselves alone the place of tertiary pinnæ, the racemes being then without leaves.—The distribution of the flowers probably affords a good character for a diagnosis. But the flowers or organs of fructifications are rarely formed and generally indistinctly preserved.

5th. From the presence or absence of free pinnules placed upon the main rachis in alternation with the secondary pinnæ. The ramification of all the species of *Archæopteris* is by opposite or sub-opposite divisions. In some species, however, the secondary branches are alternately replaced by simple pinnules and in that way the branches become alternate, though the normal sub-division is preserved, as each branch becomes opposite to a simple pinnule. That conformation appears to be peculiar to certain species and permanent in them, thus affording a reliable character. But sometimes the pinnæ are close, crowded, and it is then not possible to see if simple leaflets are interposed to the branches.

According to these remarks and considering the more important characters, the American species of *Archæopteris*, as far as I know them, may be distributed as follow:

A. Rachis smooth.

* *Pinnules alternating upon the main rachis, with secondary pinnæ.*

1. *Archæopteris hybernica*, Forbes. Pinnules generally large; borders crenulate; nervation thin, obsolete or not distinct without magnifier.

2. *A. Halliana*, Daws. Pinnules small, with a long narrow base; borders entire.—Species known to me only from description.

** *Pinnæ opposite, without alternation of pinnules.*

3. *A. denticulata*, Sp. nov. Pinnæ rigid, short; pinnules close, very oblique, narrowly obovate, denticulate at the upper border by the projection of the veins, which are excurrent terete or filiform, very distinct.

B. Rachis striate lengthwise; pinnæ opposite.

4. *A. minor*, Lesqx. Leaves of various size, entire on the borders; nervation more or less distinct; fructifications in tertiary branches or racemes, interrupting the rows of leaves in taking their place.

5. *A. sphenophyllifolia*, Sp. nov. Rachis striate lengthwise and punctulate, rough at its base; pinnæ very long; pinnules distant, very oblique, long and narrowly wedge-form, split and laciniate at the upper borders; nerves distinct.

6. *A. macilenta*, Sp. nov. Rachis striate and distantly punctulate; pinnules more open than in the preceding species, ovate, abruptly or rapidly narrowed to a short decurring pedicel, irregularly dentate from below the middle upward; nervation obsolete; flowers in long flexuous racemes at the base of the tertiary branches.

7. *A. obliqua*, Lesqx. Pinnules comparatively long, gradually narrowed to the point of attachment obliquely truncate at the apex; nerves distinct.

C. Rachis rough.

8. *A. obtusa*, Lesqx. Pinnæ long and large; pinnules of various sizes, much enlarged above from a prolonged narrow decurring base, sometimes reniform open and close together; more generally distant and oblique; nervation thin; surface smooth.

D. Rachis transversely rugose.

A. Rogersi, Daws. Rachis thick; pinnæ open, short, close; pinnules contiguous to the rachis by their upper border, entire, enlarged and rounded at the apex; nervation obsolete; flowers in opposite short racemes at the base of the tertiary branches.

To complete these remarks, which show how uncertain may be the determinations of species of *Archæopteris*, made from imperfect or too small specimens, I have to give, for supplying the want of plates, a detailed description of the new species mentioned above.

ARCHÆOPTERIS DENTICULATA, *Sp. nov.*

Texture of the plants thickish; secondary pinnæ long, linear, rigid, with a thick smooth rachis; tertiary pinnæ open or oblique, equal, short; pinnules more or less oblique, except the lowest, which is attached to the main rachis and at right angle, all narrowly obovate, obtuse, distinctly, regularly dentate at the apex.

The primary rachis is semiterete, 3–5 m. m. in diameter; the ultimate pinnæ are parallel, 7–8 m. m. distant, 2–4 c. m. long, with 10–12 pairs of small pinnules. The denticulation of the upper rounded part of the pinnules is formed by the excurrent points of the nerves, which, decurring at the narrow base of the leaves, are flabellate and dichotomous in ascending, strong and preserving their thickness even to the last divisions, in passing out of the borders as short, acute teeth.

Habitat—Rushville, Ohio, lower carboniferous.

ARCHÆOPTERIS SPHENOPHYLLIFOLIA, *Sp. nov.*

Plants large with a strong primary rugose rachis; secondary pinnæ very long, rigid, with the rachis striate lengthwise and punctulate; tertiary pinnæ oblique, rigid, linear; pinnules cuneiform, sessile, narrow, more or less deeply split at the apex; nervation distinct.

The pinnules are distant, somewhat open, exactly wedge-form at base, narrow and very much like leaflets of *Sphenophyllum angustifolium*, Germ. The aspect of the plants is quite different from that of any other species of the genus. The nervation is distinct, visible without a glass, though the surface is covered with a thin epidermis.

Habitat—Meshoppen, Pa. No. 829 of Lacoe's collection.

ARCHÆOPTERIS MACILENTA, *Sp. nov.*

Secondary pinnæ very long, oblong-lanceolate, with a striate somewhat punctulate rachis; tertiary pinnæ rigid, very oblique, not long; pinnules small, oblique, attached to the rachis by the narrow cuneiform base or a very short pedicel, irregularly dentate from below the middle upward, oval, acute or apiculate by the teeth; nerves extremely thin and close, scarcely discernible even with a strong glass; flowers in opposite long flexuous or pending racemes at the base of the tertiary pinnæ.

The secondary rachis is 5 m. m. in diameter under the point of attachment of the tertiary pinnæ, which, like the pinnules, are at an angle of 30–35°, all rigid; the pinnules are thick and distinctly denticulate from below the middle. On the large specimen figured, some of the pinnules appear truncate; but it is a mere appearance caused by the imbedding into the matter of the stone of a part of the lamina. One of the specimens, however, a fragment also figured, has all the pinnules distinctly preserved. The teeth are not produced by the prolongation of percurrent veins, as in *A. denticulata*; but by expansion, even laterally, of the thick substance of the pinnules. The fertile pinnæ or flowers are in separate opposite racemes at the base of the tertiary pinnæ, occupying the place of lower tertiary branches; they are 9–10 c. m. long on filiform, flexuous, half pending rachis.

Habitat—With the preceding, Meshoppen, in large specimens; No. 830 of Lacoe's collection.

ARCHÆOPTERIS ROGERSI, *Daws.*

"Coal Flora," p. 307, Pl. XLIX, Figs. 9, 9a.

I refer to this species a number of specimens resembling *A. minor*, but with the rachis transversely rugose, even that of the tertiary branches. The primary pinnæ are short and broad, oblong-ovate; the secondary short, half open, at an angle of divergence of 50°; pinnules small, very oblique or inclined along the rachis, 1 c. m. long, scarcely half as broad near the rounded apex. The veins though distant are rarely discernible, except at some places where they are detached by maceration. In some specimens the pinnules are very close, sub-imbricate by the borders; in others they are more distant. The aspect of the plants is peculiar, on account of the thick rachis; the short pinnæ being generally crowded and the pinnules comparatively broader, it is thus compact in all its parts. The flowers are in close short rigid racemes, occupying the place of tertiary branches at the base of the secondary ones.

Habitat—Meshoppen, Penn'a. No. 832 of Lacoe's collection.

ARCHÆOPTERIS HYBERNICA, *Forbes.*

"Coal Flora," p. 305, Pl. L, Fig. 5.

I had only before seen a single pinnule which I could consider as referable to this species. Now American specimens represent it in large pinnæ whose characters agree with the description given in "Coal Flora," l. c. The pinnules, however, are not close and sub-imbricate, but rather distant, variable in size from 2–3 c. m. long and 8–16 m. m. broad, near the apex; they are oblique or inclined but not decumbent upon the rachis, being attached to it by a short decurrent pedicel; the upper borders of the pinnules are distinctly crenulate, even more or less deeply split. The fruiting racemes are placed in the middle of the tertiary branches, as in *A. minor*, "Coal Fl." Pl. L, Fig. 3.

Habitat—Meshoppen. No. 697 of Mr. R. D. Lacoe's collection.

LYCOPODIACEÆ.

"Coal Flora, p. 355.

LYCOPODITES, *Brgt.*

Under the generic name of *Lycopydites*, Goldenberg has published a number of species of coal plants which, he says, he considers as true Lycopods of the ancient world, they having the same characters as the plants now described under the generic names of *Lycopodium* or *Selaginella*. These species are, therefore, rightly described by Schimper under the name of *Lycopodium*. They are herbaceous plants with homomorphous or dimorphous leaves, bearing fructifications in cylindrical spikes, like living species, their spores, says Schimper, being still unknown.

From what we have observed in the American coal measures, some species representing Lycopods by their branches, their leaves and their ramification are not herbaceous. They cannot be referable to the genus *Lycopodites*, as defined by Goldenberg, and nevertheless they are not branches of *Lepidodendron*, though they have been considered as such by Brongniart and other authors. Brongniart says, "Tableau des genres," p. 40, that the plants, really analogous to Lycopods, are very rare in a fossil state and that most of those which he himself or other authors had referred to *Lycopodites* are probably young branches of *Lepidodendron* or of Conifers. I have already published some species of this kind, apparently contradicting the opinion of Brongniart, and those we have now to add show still more evidently that some plants of the Coal Flora, closely allied to Lycopods, differing from those of our epoch by woody stems, are not referable to *Lepidodendron*. The definition of the genus *Lycopodites* may therefore be preserved as quoted from Goldenberg, "Coal Flora," p. 357, with the simple modification of "Plants herbaceous or woody."

Until now, no very large woody plants have been found as representatives of this genus. The stem, Pl. CVI and the strobile, Pl. CVII, Fig. 1, seem to indicate that these fragments are mere parts of plants of a far greater size than are any of the species of Lycopods living at the present epoch.

LYCOPODITES ARBORESCENS, *Sp. nov., Pl. CVI, Fig. 1.*

Stem thick, woody; primary branches derived nearly at right angles and diverging by an inside curve, dichotomously sub-divided; branchlets flexuous or pending; leaves oblong, acute, concave, without medial nerves, and imbricated in spiral order.

The stem, nearly 3 c. m. in diameter, is flattened, irregularly punctulate, transversely, distantly rugose and marked longitudinally by distant irregularly straight or flexuous striæ. It has no remains of leaves or scales and no trace of scars left as point of attachment of the leaves. The branch, nearly one c. m. in diameter at the point where it separates from the stem, is first smooth, or without leaves, gradually slightly narrowing to the point where it forks into two divisions of equal thickness, 4 m. m. in diameter, which are covered with distant compressed scalyform leaves, to the point where the branches, forking again, become flexuous pending and then covered with imbricated half-open leaves, 5 m. m. long and 1 m. m. broad in the middle.

No point of comparison is found for this species in any of the European authors who have described the plants of the coal. But for the leaves, at least, and their dispositions, the analogy is clearly marked with *Lycopodites cavifolius*, Lesqx., described in "Rept. of the Geol. Survey of Ky.," by D. D. Owen, p. 437, Pl. ined; and also with *Selaginites* (*Lycopodites*) *crassus*, Lesqx., "Geol. Rept. of Ill.," II, p. 446, Pl. XXXIX, Fig. 8. In both these species, which are perhaps varieties of the same, the leaves have identical characters in their concave shape and the total absence of a middle nerve.

Though no living species can be compared to that figured Pl. CVI, analogy of ramification is remarked in some Lycopods of the present epoch, for example, in *Lycopodium inflexum*, Sw., whose primary stems larger than the secondary, are horizontal in their direction, while the secondary curve upward in the same way and the same angle of divergence as that remarked upon the fossil species, of which the stem was possibly creeping or growing horizontally, like that of the above named plant.

Habitat—Subconglomerate measures of Arkansas. Communicated by F. L. Harvey. No. 257 of Lacoe's collection.

LYCOPODITES FLEXIFOLIUS, *Sp. nov.*, *Pl. CVI, Figs. 3, 4.*

Dichotomous from the base of the largest branches; primary stipe irregularly, interruptedly costate, marked by punctiform scars of leaves; branches long; leaves distant, in spiral order, oblong, obtuse, narrowed to the point of attachment, inflated in the middle, but not distinctly nerved, all reflexed or horizontally recurved.

The primary or lowest part of the stem, Fig. 4, is about 1 c. m. in diameter, rugose, costate lengthwise, punctulate by irregular round dots, points of attachment of the leaves; the stems, gradually diminishing in size by forking, are very slender, scarcely 1 m. m. in diameter in their ultimate divisions. The leaves are generally partly embedded into the stone in such a way that their real form is not easily ascertained. They appear oval or oblong, obtuse, generally truncate, as partly embedded, all reflexed and distant, and without distinct costa, though they are generally inflated in the middle.

Habitat—Subconglomerate coal measures of Arkansas. F. L. Harvey.

LYCOPODITES SIMPLEX, *Pl. CVI, Fig. 2.*

Spikes long, linear; sporanges and leaves in horizontal close rows; sporanges small, apparently round, axillary; leaves short, small, narrowly lanceolate-acuminate.

This fragment evidently represents the fruiting spike of a species of this genus, being analogous in its characters to the spikes of living species of Lycopods, *Lycopodium inflexum*, for example, already quoted, and especially very similar to *Lycopodites leptostachys*, Goldenb. "Fl. Sarræp." I, p. 12, Pl. 1, Fig. 4, of the Carboniferous of Germany. In this new species, the rows of sporanges are close, very small, round, each apparently in the axil of a small leaf or scale, these leaves being open-erect, narrowly lanceolate, sharply acuminate, 2 to 3 mill. long, less than 1 m. m. broad at base. The fruiting spike appears to have been very long, the fragment preserved being still 14 c. m. long, cylindrical, very gradually narrowed upward, being at the broken base 4 m. m. in diameter (flattened), and 3 m. m. at the apex.

Habitat—Subconglomerate Campbell's Ledge, Pittston. Lacoe's No. 258.

LYCOPODITES LACOEI, *Sp. nov., Pl. CVII, Fig. 1.*

Lepidostrobus, Lacoei, Lesqx., "Coal Flora," p. 439.

Spike long and broad, bearing crowded round tubercles or sporanges in horizontal rows, axillary or interspersed with linear-lanceolate leaves; pedicel long, slightly flexuous, enlarged to its base and upward to the base of the spike.

The specimen being perfect the figure represents the whole organism as far as it is discernible. The spike without the pedicel is 34 c. m. long, 1½ c. m. in diameter in the middle, being gradually narrower to the apex, where it is only half as broad. The pedicel is 13½ c. m. long, 5 m. m. in diameter in the middle where it is narrower. The sporanges are globular, about 2 m. m. in diameter, sometimes compressed or irregular in shape, on account of their crowded position in close horizontal rows. The scales or leaves, about 1 c. m. long, a little more than 1 m. m. broad at base, are erect, or inclined upward, lanceolate-acuminate, without distinct medial nerves, either placed between or at the base of the sporanges, which, however, do not appear axillary. The long pedicel, gradually narrowed to the middle, is gradually enlarged to its base in the same proportion. This long, quite smooth pedicel seems like an anomaly in the fructification of Lycopods. But, in fragments of other specimens of this species, the pedicel has been found chaffy or scaly at its base, and gradually covered above by distant lanceolate imbricate leaves, as seen in the description of *Lepidostrobus*, *Lacoei*, l. c., to which this species was first ascribed. The characters of this spike are far different from those of a *Lepodistrobus*, as can be seen by the specimen figured.

Habitat—Olyphant, No. 1 vein. Collection of Mr. R. D. Lacoe, No. 524.

LEPIDOPHLOIOS, *Sternb.*

Lomatophloios, Corda, "Coal Flora," p. 418-429.

The character of the plants described as *Lomatophloios* by Corda have been carefully studied by the author, who has given sufficient information on their internal structure to prove their relation to the *Lycopodiaceæ*. Less, however, is known of the outside characters of these plants as possibly recognized in fragments of stems or leaves, like those found preserved in the carboniferous of this continent, and especially on their fructification. Plate CV, therefore, has its value, in representing altogether the best specimens obtained from the American Coal measures; fragments of bark, with the scars of leaves upon small and large branches, either corticated or decorticated, and parts of a strobile bearing spores and blades or bracts of sporanges, showing the enormous size of the cones pertaining as fructifications to this genus.

LEPIDOPHLOIOS DILATATUS, *Sp. nov.*, *Pl. CV, Figs. 1-4.*

Scars of bolsters transversely much enlarged, contracted and acute at the sides, rounded in the upper and lower parts, mamillate in the middle; leaf scars narrowly rhomboidal, transversely enlarged; lateral angles narrowly acute, the upper and lower very obtuse; vascular scars three, the middle larger. Cones or sporophores very large, composed of appressed linear large scales, cordate truncate at base, covering large and long agglomerations of macrospores.

Possibly this species may be the same as that described and figured "Coal Flora," p. 424, as *Lepidophloios macro-*

lepidotus, Goldenb. It seems however, different, by the bolsters and leaf-scars much enlarged sidewise, with very narrow acute lateral angles. In some specimens of *L. laricinus*, the bolsters and leaf-scars are often much enlarged laterally; but even in specimens of smaller size and far more distinctly so in those with large scars, the bolsters are comparatively longer lengthwise than transversely. In Goldenb. "Fl. Sarræp.," Pl. XVI, Figs. 1, 6, the small bolsters measure 5 m. m. in both directions, the large ones 2½ c. m. transversely and about 3 c. m. vertically. In the American species the small scars more than 1 c. m. in lateral width are scarcely 5 m. m. in vertical direction, and the large ones preserving the same proportions are 3½ c. m. transversely and scarcely 1½ vertically measured. There is also a marked difference in the size and form of the scars of the decorticated surface. In *L. laricinus* the sub-cortical scars are round, very small, while in the American species they are oblong or narrowly oval, more than twice as long as broad.

The cone, Pl. CV, Fig. 1, appears to be of the same kind as the fragment in "Coal Flora," p. 427, Pl. LXVIII, Fig. 6, mentioned as representing fructifications of *Lepidophloios*. The figured part appears to be a flattened fragment of the outside of a large cone with imbricated scales or bracts, covering long linear agglomerations of macrospores which are of the same form as those figured "Coal Fl.," Pl. LXVIII, Figs. 7 7*b*, only a little longer. At the upper part of the specimen, the blades have been destroyed and have left distinctly exposed the agglomerations of spores compressed into long cylindrical masses, 3 to 4 c. m. long, 1 c. m. broad. Flattened layers or lumps of these spores are easily separated from the stone whereupon they are compressed into a thickness of 2 to 3 m. m. and contain an innumerable number of spores without trace of any sporanges. The blades seen on the side of the specimen are most of them broken at the apex, the few remaining upon the surface appear like fragments or impressions of imbricated bracts covering the spores or perhaps their sporanges, of which, as said above, no trace is seen in connection with the clusters of spores.

Habitat—The specimens figured are all from Cannelton, Penna., communicated by I. F. Mansfield. Others of the same kind, representing stems and branches, but no fructifications, have been sent by Dr. Britts from Clinton, Mo.

LEPIDODENDRON.

"Coal Flora," p. 363.

LEPIDODENDRON STERNBERGII, *Brgt., with cone, Pl. CVII, Fig. 2.*

"Coal Flora," p. 366.

LEPIDOSTROBUS VARIABILIS, *Ll. and Hutt., Pl. CVII, Fig. 2.*

"Coal Flora," p. 434.

Branch dichotomous covered with indistinct small regularly broadly rhomboidal or square bolsters, bearing a cylindrical obtuse comparatively long, strobile.

This beautiful specimen shows part of a dichotomous twice forking branch, 2 c. m. in diameter in its lower undivided part, reduced first to 1 c. m. after the first forking, and to 5 m. m. in the lateral branchlet bearing the cone. The scars upon the stems have only the outlines preserved. Their shape is identical to that of the scars of *L. Sternbergii*, Brgt., or *L. dichotomum*, St., as figured "Fl. d. Vorw.," Pl. 1, in large stems with branches covered with scars, like those of our specimen.

The cones of this species, as represented by Schimper, "Paleont., Veget., Pl. LVIII, Fig. 5, are only a little smaller, cylindrical, of the same form as that of Fig. 2, and most probably, as Schimper remarks, it is to this *Lepidodendron* that are referable the cones described as *Lepidostrobus variabilis* in Ll. and Hutt., "Foss. Fl." Pl. X and XI, and those figured by Brgt., "Veget. Foss.," Vol. II, Pl. 22 and 23. All these strobiles are cylindrical, variable in length from 5 to 15 c. m. long and from 1½ to 2 c. m. in diameter. Though the blades of the sporanges are not distinctly preserved upon the American specimens, they appear from their impressions upon the strobile to have been short, narrowly lanceolate, like those seen upon the cones figured by Lindly.

Habitat—Braceville, Grundy co., Ill. No. 601[e] of Lacoe's collection.

LEPIDODENDRON (BERGERIA) MARGINATUM, *Presl. Pl. CVII, Fig. 3.*

Bergeria marginata, Presl. in St., "Fl. d. Vorw.," p. 134, Pl. LVIII, Fig. 16.

Bolsters rhomboidal-ovate, narrowly margined all around, acute at the base and the apex, straight; marked near the upper margin by a globose tubercle, and in the middle by a short vertical line.

The specimen is concordant in all its details with the figure given by Sternberg, l. c. The bolsters are 12 to 14 m. m. long, 10 to 12 broad in the middle.

Habitat—Clinton, Mo., communicated by Dr. J. H. Britts; Plymouth, Pa. F vein. No. 357, Lacoe.

LEPIDOSTROBUS, *Brgt.*

"Coal Flora," p. 431.

LEPIDOSTROBUS MACROCYSTIS TRUNCATUS, *Lesqx., Pl. CVIII, Fig. 1.*

Lepidophyllum truncatum, Lesqx., "Coal Flora," p. 458, Pl. LXIX, Figs. 9, 10.

The specimen represents a somewhat large axis, flattened by compression, bearing large oblong sporanges truncate at the point of attachment, obtuse at the apex, sometimes a little contracted in the middle, smooth and transversely wrinkled on the surface, filled with macrospores which are distinctly seen without the glass, the spores being 1 m. m. in diameter.

I have three specimens of different fragments most alike. The lower sporanges, attached by the base, are without sporangiophores, thick, a little convex on the surface, apparently filled with spores. The upper ones are open, and the spores are still in groups upon them, though the shale is covered with a profusion of those which have been strewn around. The two figures 9 and 10 of Pl. LXIX show two sporanges, one empty of its spores, the other with some of them still attached to it. The spores are represented, Fig. 9, in their natural size and Fig. 9*a* enlarged. The relation of this spike with that described as *Lepidostrobus macrocystis*, Pl. LXIX, Fig. 1, is easily recognized.

It might be supposed that these large *Lepidostrobi* are fructified cones of *Lepidophloios*. The one represented Pl. CIV is more evidently that of a *Lepidophloios* and is of a far different character. They might be referable to some kind of *Sigillaria*. It is a remarkable fact that though fruiting remains of *Lycopodiaceæ* abound at Cannelton, species of *Lepidodendron* are there comparatively rare.

Habitat—Cannelton. Sent by I. F. Mansfield.

LEPIDOPHYLLUM, *Brgt.*

"*Coal Flora*," *p. 447.*

LEPIDOPHYLLUM CULTRIFORME, *Sp. nov., Pl. CVII, Figs. 13, 14, CVIII, Fig. 2.*

Sporanges large, quadrangular, oblong, sometimes shorter, nearly square; blades enlarged at base, long, acuminate from the middle; medial nerve narrow.

This species is generally found in separate sporanges with their blades still attached to them. The sporanges vary from 10 to 15 m. m. long, being generally 7 m. m. broad, flattened; the blades, generally open near the base, curved up from the middle, are 2 c. m. long and 5 m. m. broad at the part where they pass out of the cone. I have not seen any of the sporanges open, but some have the surface marked with points or impressions of the spores underneath.

The form, Pl. CVII, Figs. 13, 14, is slightly different. The sporanges are narrower, not as distinctly quadrangular, obtuse at both ends; the blades are broader, generally shorter, varying in length from 1½ to 2 c. m. and being 3 m. m. broad at the base.

Habitat—The first form is not rare at Cannelton; the

last has been collected in numerous specimens in the sub-conglomerate, Campbell's Ledge near Pittston, by Mr. Lacoe.

LEPIDOPHYLLUM FALLAX, *Sp. nov.*, *Pl. CVII, Figs. 4, 5.*

Size large; blades linear or oblong-lanceolate, acute or acuminate; medial nerve broad; sporanges comparatively short, narrowed to a blunt point.

The species is intermediate in its size and form between *L. acuminatum* and *L. lanceolatum*. The blades average 5 c. m. long and about 1 c. m. broad in the middle; the sporanges a little more than 1 c. m. long, are comparatively longer than those of *L. lanceolatum* and shorter than of *L. acuminatum*. Fig. 4 may represent a different species. The blade is generally narrower, more linear, less acuminate, the medial nerve is not half as thick, but the sporange is of the same form and character.

Habitat—Specimen Fig. 4 is from Rhode Island anthracite coal, that of Fig. 5 from Cannelton, Pa. Both are in the collection of Mr. Lacoe as No. 892 and 892*a*.

LEPIDOPHYLLUM CAMPBELLIANUM, *Sp. nov.*, *Pl. CVII, Figs. 6, 7.*

Blades of medium size, gradually enlarged from a narrow base to far above the middle, then rapidly acuminate; sporanges small obconical, acute at base.

The blades are 5 to 6 c. m. long, 5 m. m. broad at the point of union to the sporanges, gradually enlarged above to below the apex, where they are 10 to 15 m. m. broad, and then rapidly narrowed to a sharp point or acumen. The sporanges are only 5 to 6 m. m. long. The medial nerve is comparatively thick, 2 to 3 m. m. at base.

Habitat—Subconglomerate, Campbell's Ledge, Pittston. No. 658 of Lacoe's collection.

LEPIDOPHYLLUM GRACILE, *Sp. nov.*, *Pl. CVII, Fig. 8.*

Size small; blade exactly narrowly lanceolate, sharply pointed; sporanges comparatively long, narrowed to a very acute base.

The blade, 2 c. m. long, is only 4 m. m. broad at the point of union to the sporange, which is 6 m. m. long; the nerve is broad.

Habitat—Same as last above. No. 894, collection of R. D. Lacoe.

LEPIDOPHYLLUM MINUTUM, *Sp. nov., Pl. CVII, Fig. 9.*

Blade broad and short, a little longer than broad, 5 m. m. long, 4 m. m. broad at base or triangular, blunt at the apex, with a thick medial nerve.

The blade only has been seen. It is shorter than that of *L. brevifolium* in "Geol. of Penna.," p. 876, Pl. XVII, Fig. 6. But probably it is referable to the same species. As the sporange is destroyed, identification is impossible.

Habitat—Archibald, Pa. No. 673 of Lacoe.

LEPIDOPHYLLUM CORIACEUM, *Sp. nov., Pl. CVII, Fig. 10.*

Blade oblong-lanceolate, acuminate, broader at and above the base, of thick texture, polished, broadly nerved and marked on each side of the costa by two thick inflated striæ parallel to the borders, passing from the upper part of the blade and vanishing above the base; sporange short, as broad as the blade at the point of union, constricted into an acuminate base.

The blade, 12 m. m. broad at the base, nearly 4½ c. m. long, is narrowed at the apex to a sharply apiculate po nt; the space between the striæ is about equal to the width of the costa, 2 m. m.; the striæ are not sharply defined, but obtuse, the space between them being somewhat concave. The sporange is 1 c. m. long.

Habitat—Cannêlton, Penna., rare. No. 672 of Lacoe's collection.

LEPIDOPHYLLUM ELEGANS, *Sp. nov., Pl. CVII, Fig. 11.*

Blade short, slightly enlarged in joining the sporange, contracted above the base, broader above the middle, then tapering up to a sharp point; sporange slightly contracted and narrowed to an acuminate base.

The blade is 18 m. m. long, 7 m. m. broad above the point

of connection to the sporange, and 1 c. m. at the middle; the sporange is 6 m. m. long and as broad, a little narrower than the blade at the line of connection; the costa is comparatively thick at the base of the blade.

Habitat—Brown colliery, Pittston. Lacoe, No. 737.

LEPIDOPHYLLUM TUMIDUM, *Lesqx.*, *Pl. CVII, Fig. 12.*

"Coal Flora," p. 448.

The only remark to be made on the description given in the Flora, l. c., is that the blade is not acuminate but only pointed, nor carinate by the thick midrib but tumescent along it.

TÆNIOPHYLLUM, *Lesqx.*

"Coal Flora," p. 461.

I have nothing to add to the description of the genus, l. c. The description of the following species merely confirms the supposed relation of the plants of this kind to the *Lycopodiaceæ.*

TÆNIOPHYLLUM BREVIFOLIUM, *Sp. nov.*, *Pl. CVIII, Figs. 3, 3a.*

Leaves short, closely appressed, flat by compression, but apparently tubular and hollow in their original state; macrospores spread in flakes upon the surface along the leaves, seemingly coming out from the leaves as from long cylindrical sporanges containing them.

There are two fine specimens of this species. The leaves are compressed one upon another, without symmetrical order, short, about 3 c. m. long, 3 m. m. in diameter, linear, narrowed into a blunt apex. The bundles of spores are irregularly placed, some in small compact groups, others more disseminated or spread along some leaves as if forced by compression out of a tubular envelope or sporange. All the spores are macrospores, 1 m. m. in diameter, as figured f. 3*a*. The fragments show the plant in its whole length, as an agglomeration of leaves compressed from the base to the apex which is genrally broken. The texture, as seen at the surface is that of *Tæniophyllum contextum*, Lesqx., "Coal Flora," p. 465, Pl. LXXXII, Figs. 2, 2*a*. These plants may be the young shoots of the same species. In the description of *T. contextum*, I remarked on the relation of the plants to *Isoetes*. The relation is still more evident in these short plants. But the nature and the character of the sporanges are as yet unexplained.

Habitat—Cannelton. No. 774 of Lacoe's collection.

TROCHOPHYLLUM, *Lesqx.*

"Coal Flora," p. 63.

That generic name was used, l. c., for the provisory description of vegetable remains which, in a too deficient state of preservation, could not be definitively described, and which, therefore, could not be affixed to their legitimate place until their affinities had been demonstrated by the discovery of better specimens.

The first of the two species placed in the genus *Trochophyllum*, viz: *T. clavatum*, described l. c., p. 65, Pl. III, Figs. 21–23*a*, has been recognized as representing half destroyed fragments of strobiles of *Sigillaria*, described here below as *Sigillariostrobus Laurencianus*. Of the other species, *T. lineare*, l. c. p. 64, Pl. III, Figs. 24–25*b*, I have now received a large number of specimens which demonstrate the affinities of the plant with the *Lycopodiaceæ*. Fragments of the same kind, though of different aspect, having been described by Prof. Dawson under the name of *Ptilophiton*, this last generic name should be admitted, while that of *Trochophyllum*, for as far as it refers to the plants I have described in it, has to be abandoned.

PTILOPHYTON, *Daws.*

Notes on Scottish Devonian Plants, 1878.

Branching plants; branches bearing long slender leaves in two or more ranks, giving them a feathered appearance; venation circinate. (Dawson.)

This definition is limited to one species only of the genus, which was then still imperfectly known. The numerous specimens of different facies, mostly of the same type as that represented in Coal Flora, Pl. III, Fig. 24-25*b*, which have been discovered in Ohio, necessitate a modification of the primitive generic description, as follow :

Branches short, apparently in loose tufts, mostly simple, generally erect or circinate when young ; rachis narrow, transversely tuberculate; leaves or appendages distichous or in whorls, rigid, oblique, either very thin, filiform, or larger, linear, obtuse, rounded at base to their point of attachment to a tubercle, vesicular and flattened; sporanges disposed upon the stems in the same way as the leaves, but longer, distinctly tubulose-triquetre; either empty or filled with macrospores.

PTILOPHYTOM VANUXEMI, *Dawson.*

Quat. Journ. Geol. Soc. XVIII, p. 314, Pl. XVII, Fig. 57.

Filicites, Hall, Geol. Rep. of New York, p. 273, Fig. 125; Vanux. Rep. p. 175, Fig. 46. Plumulina plumaria, Hall, 1858.

Stem slender, simple, rarely forking ; leaves pinnate, contiguous, linear, a little more than 1 c. m. long, apparently cylindrical and tubular, narrowed to the point of attachment, oblique.

Habitat—Chemung group of New York, Hall.

I know this species only by the figures and descriptions of the authors.

PTILOPHYTOM GRACILE.

Plumulina gracilis, Shumard.

Stems short, erect or circinate above; rachis comparatively thick ; leaves or appendages very thin, strict, contiguous, filiform, oblique.

I cannot say positively that the specimens of Ohio represent the same species as that of Shumard, as I have not had an opportunity of seeing its decription or figures. But Hall in a Note on the genus *Plumulina*, "13th Rep. of the State Museum of New York," April, 1879, remarks, that regarding his species *P. plumaria* as the type of the genus, he recognized that of Shumard as *P. gracilis.*

In my specimens, the leaves are as thin as fine hairs, not discernible to the eyes, very close, oblique, parallel, 5 to 6

m. m. long; the rachis, however, is comparatively broad, 1 m. m. or as broad as is the small pinnæ of the following species.

Habitat—Found in a large boulder or ferruginous concretion in beds of clay of the Waverly Sandstone near Rushville, Ohio.

PTILOPHYTOM LINEARE, *Dawson*.

Fossil plants of the Devonian and Upper Silurian of Canada II, p. 123, Fig. 5, woodcut.

Trochophyllum lineare, Lesqx., "Coal Flora," p. 64, Pl. III, Figs. 24, 25b.

Branches simple with a slender articulate and tuberculate rachis; leaves either short and small, or longer, linear, obtuse, rounded and narrowed at base to the point of attachment upon the tubercles which appear in close transverse rows of three upon the upper flattened surface, flat, or tumid and vesicular, black, coriaceous, irregularly wrinkled on the surface by compression; sporanges cylindrical-quadrate and tubular, empty or filled with macrospores.

The species is represented under divers aspects. 1st. In very small flat thick or vesicular leaves 2 to 15 m. m. long, 1 to 3 m. m. broad, evidently of vegetable nature, as shown by the coaly texture and black color. 2d. In long narrow cylindrical leaves 1–2 m. m. broad, 1½ to 2½ c. m. long, of the same texture as those of the preceding group. 3d. In cylindrical-quadrate tubular appendages, apparently sporanges, containing large macrospores and showing a quadrangular transverse section. The rachis and the point of attachment of all these leaves or sporanges are of the same character.

I have been long time in doubt on the true nature of those tubular appendages which I now consider as sporanges. For though the matter embedding the fragments of plants, a very hard concretion of ferruginous clay, bears generally an abundance of large spores spread all around the branches and the leaves, I was unable to distinctly see the spores in the sporanges which are sometimes crushed by compression. I have now two specimens which show groups of spores enclosed in the sporanges. They have been carefully figured with the fragments of these remarkable plants for future elucidation of their characters. They evidence the nature of these organs as vegetable, and show their relation to be with the *Lycopodiaceæ*. I am not yet certain that the leaves are distichous. As they are attached by a very slender base to the tubercles of the stems and are apparently very brittle, the lateral ones only may have been preserved, while those of the upper part were broken and destroyed. The tuberculate scars are in rows of six, apparently; at least, three are exposed upon the upper surface of the flattened rachis and therefore the leaves appear to have been verticillate like the sporanges of *Sigillariostrobus Laurencianus*, which, upon some of the specimens, are exposed to view by the deterioration of the surface and appear flattened on both sides of the rachis. The rachis, however, is marked with the scars of the points of attachment upon its upper surface, very much like the stems of *P. lineare*. This is clearly seen in comparing in "Coal Flora," Pl. III, Figs. 21–23*a* with Figs. 24–25*b*, the first being an axis of *Sigillariostrobus*, the second, the rachis of *P. lineare*.

The relation of these plants to vegetables of the present epoch is unknown. They were apparently living in low marine water; for to their stems, even to the leaves, are sometimes attached small bud-like protuberances appearing like very small shells or their eggs. These are considered by Hall and Dawson as small brachiopods.

Habitat—Near Rushville, Ohio, in the same boulder and in numerous specimens with only two of the preceding species. All the specimens are preserved in Mr. Lacoe's cabinet.

SIGILLARIEÆ.

"Coal Flora" 2, p. 466.

Nothing more positive is known at the present time on the relation of the *Sigillarieæ*, either to the *Lycopodiaceæ* as cryptogamous plants, or to the *Cycadeæ* as Gymnosperms, than when the first part of the Coal Flora was published. That question is now reconsidered by two celebrated anatomists and physiologists, Professor M. B. Renalt, of the Museum of Natural History of Paris, who has admitted Brongniart's views, relating the *Sigillarieæ* to the *Cycadeæ*, and Professor W. C. Williamson, of Manchester, who sustains the contrary opinion. In the controversy which has been extremely interesting to phytopaleontologists, both authors have produced, in support of their ideas, remarkable works on the texture of *Sigillaria* and *Lepidodendron;* nevertheless, the question remains undecided or just at the same point as it was when first considered by Brongniart. The little I have said on the subject, Coal Flora, l. c., is sufficient to show why I am disposed to admit the relation of *Sigillaria* to *Lepidodendron*, and I should not come again to the subject if I had not recently obtained some new evidence in favor of that opinion. That is explained in the description of the two following species of *Sigillaria*, one only of which is new. To my regret, the specimens have been obtained too late to be figured in this volume.

SIGILLARIA, *Brgt.*

"Coal Flora," p. 467.

SIGILLARIA MONOSTIGMA, *Lesqx.*

"Coal Flora," p. 468, Pl. LXXIII, Figs. 3–6.

The new specimen represents a fragment of stem or branch with the cortex marked with the areoles and the regular rugosity of surface of *Sigillaria monostigma*, and part of an ear or strobile, *Sigillariostrobus*, referable to the same species by its association and its position partly under the branch upon the same fragment of shale.

The branch, 2½ c. m. broad, has still some leaves attached to the scars. The areoles are closer than on the specimen Pl. LXXIII, Fig. 3, only 4 to 5 m. m. distant; but they are of the same size and form, marked by a single central vascular scar, and the surface between the areoles is striate in the same way by lines tending obliquely and in opposite direction toward the scars. The leaves of the species, which I had not seen before, are somewhat broad, 3 m. m. in diameter, slightly keeled in the middle, on the lower face, by a thick medial nerve, which is concave on the upper. There is, also, between the medial nerve and the borders a thin veinlet distinct here and there upon the leaves, none of which are preserved whole. The strobile, partly covered by the stem, is long and linear, 2 c. m. broad, flattened, covered with imbricate enlarged rhomboidal scales, tumid at base and closely appressed. There are still two other detached fragments of the same species, if not of the same strobile, both linear, each 8 to 9 c. m. long. One is covered the half of its length by the imbricated scales which, destroyed on the other half, leave exposed to view the sporanges attached in horizontal rows to an axis, 3 to 4 m. m. broad. The sporanges are nearly at right angles, oblong, vesicular and flattened by compression, 5 to 6 m. m. long, 2 m. m. broad. The third specimen is, by the destruction of its covering of scales and part of the sporanges, open its whole length. In the lower part, the sporanges have been already detached, but the sporangiophores are still in place, fixed at right angles to the rachis, like filaments very narrow toward the base, enlarged or clavate outward, indeed, exactly of the same form and in the same position as what I considered to be the leaves of *Trochophyllum clavatum*, "Coal Flora," p. 65, Pl. III, Fig. 21. As the scars of the points of attachment of these sporangiophores are also of the same nature and in the same position as in Fig. 22 and 23, l. c., these as yet problematic remains, provisionally placed in the genus *Trochophyllum*, are now recognized as fragments of cones of *Sigillariostrobus* whose sporanges have been detached by maceration, the sporangiophores being still partly left in place, but often already partly broken, as seen Fig. 21, l. c.

These spikes, which may be named *Sigillariostrobus Laurencianus*, though very different from any kind of *Lepidostrobus*, have the same essential conformation, viz: sporanges about of the same form, attached in horizontal rows to a vertical axis, supported by persistent sporangiophores with lanceolate scales, turned up and imbricate.

Hence, if *Sigillaria monostigma* should be considered a true species of this genus, the relation of the *Sigillarieæ* to the *Lycopodiaceæ* would be proved. This relation, however, is contested in regard to the few species which, like *S. monostigma*, have only one vascular scar marked upon the areoles. But the question may be more clearly examined after the description of the following.

Habitat—The fragments described above, No. 464 and 464*a* of Lacoe's collection, have all been found near Lawrence, Kansas, in a bed of soft sandstone of the Carboniferous.

SIGILLARIA GRAND'EURYI, *Sp. nov.*

Supercortical areoles not distant, rhomboidal, constricted and acute at both sides, rounded at the upper and lower border, symmetrical, convex, traversed from one side to the other by an inflated line marked in the middle by a single round vascular scar; surface between the scars smooth; decorticated surface also smooth, the outlines of the leaf-scars being totally effaced, and the vascular scar only marked by a very distinct small oval concave areole.

The species closely resembles *S. monostigma*, from which it differs by the less distant scars, only 2 m. m. from each other, the space between them smooth, the leaf scars tumid, somewhat larger, transversely crossed by a narrow ridge or terete line, and the shape of the subcortical areoles represented by small oval scars concave inside. The specimen has the leaves still attached to the areoles, some of them at least. They are somewhat enlarged at and toward the base, 3 to 4 m. m. broad at their point of attachment gradually narrower to 2 m. m. up to the apex or in their whole preserved length, the longest measuring 12 c. m. The medial nerve is broad, enlarging like the leaves toward the base, where generally one thin vein is seen on both sides of the medial one. The surface is opaque.

The specimen which is large shows the cortical surface with the leaves spreading along it upon the left side, either separately or by a few together or crowded into a compact flattened mass; while on the other side, along the cortical part and bordering it, there is a well preserved axis of *Artisia* which seems to have been the pith of the trunk or branch whose bark or cellular substance is flattened along it. The connection of the two parts is not ascertainable but appears as distinct as that of the leaves to the scars toward which they are directed, though most of them have been detached. As it is well known, those *Artisia* have been considered by Corda as the central axis of *Lepidophloios*, and by more recent authors, as that of *Cordaites*, also. Hence, the connection of that axis to a stem of *Sigillaria* would relate the species to *Lepidophloios* or to the *Lycopodiaceæ*, as nothing in the appreciable character of the vegetable fragment has any kind of analogy to *Cordaites*.

This relation is admitted by Grand'Eury, but only for the species of *Sigillaria*, which, like the two described here, have the vascular scar marked by a single round point. But that author does not consider them as true *Sigillaria*, and separates them in a peculiar genus under the name of *Pseudo-Sigillaria*, a genus which he places as intermediate between the *Lepidodendreæ* and the *Sigillarieæ*. He remarks that the affinity to the *Lepidodendreæ* is especially evident in the decurrence of the subcortical scars forming longitudinal marks, either splits or lineal fusiform tubercles, while in *Sigillaria* the subcortical scars are not elongated by decurrence, but flattened, even depressed. In the two species of *Sigillaria* described above, the first, according to the remark of Grand'Eury should be referable to the *Lepidodendreæ* on account of its elongated tubercles and splits, as seen Pl. LXIII, Figs. 5 and 6; while the second should go to the *Sigillarieæ* in consideration of its small oval depressed or concave subcortical scars. The character of the form of the subcorticated leaf-scars cannot, therefore, be considered as indicating the relation to the *Lepidodendreæ* or to the *Sigillarieæ*.

Considering the question on another point of view, it is evident that the spikes or strobiles found with *Sigillaria monostigma* are of the same character as those figured by Goldenberg, "Fl. Sarræp. foss.," Pl. IV, Fig. 3, and Pl. X,

Fig. 2, and if, as the author thinks, these spikes are the fructifications of *Sigillaria*, the reference of *Sigillaria* to the *Lycopodiaceæ* is a matter of course. But on this subject again, Grand'Eury says, l. c. p. 159, that the spikes which he has found attached distinctly to stems of *Sigillaria* differ perhaps entirely from those considered by Goldenberg as the fructifications of *Sigillaria*. In confirmation of that, he represents l. c. Pl. XIV, Fig. 4, a long spike, curved at base, with close verticils of leaves open at their base, then curved upward and loosely imbricate, somewhat like the ears represented Pl. LXXXIX, Figs. 1, 2, organs which I have always considered as related to the *Calamarieæ* like the other species of *Calamostachys* and *Volkmannia*, but which the French author would be disposed to admit as fruiting organs of *Sigillarieæ*.

According to the views which I have had until now on the relation of *Sigillaria* and *Lepidodendron* and on the characters of the divers organs pertaining to species of this genus, I can but admit, from the nature of the two species considered above: 1st. That *S. monostigma*, by its fruiting *Sigillariostrobus* has its relation to the *Lycopodiaceæ*. 2nd. That also *S. Grand'Euryi*, with an *Artisia* for the pith of its stem is related to the *Lepidophloios* and therefore also to the *Lycopodiaceæ*. 3d. That about the reference of these two species to the genus *Sigillaria*, it may be contested as it has been by Grand'Eury; but judging from the characters appreciable to the sight, the internal structure being unknown, they cannot be admitted in the genus *Lepidodendron*, and their place and affinities are with *Sigillaria*.

Habitat—Cannelton. Lacoe, No. 735.

SIGILLARIA BRARDII, *Brgt.*

"Coal Flora," p. 477.

A splendid specimen of that species sent to me for examination by Mr. Sam. Huston, of Richmond, Ohio, shows a peculiar character not as yet remarked upon any other American specimens of the genus. It is a young trunk 36 c. m. long, with a circumference of 23 c. m. at base, 20 c.

m. at the broken top, transversely marked with four inflated circles or knots resembling articulations, at nearly equal distances from each other, the lowest internode being 10 c. m., the middle 9½, the upper 9 long. The internodes are largest at their base or nearest the articulations. The lowest, 23 c. m. in diameter above the 1st articulation, is 22½ c. m. under the 2nd; the middle, 23 c. m. above the 2nd articulation is nearly the same under the 3rd; the 3rd articulation has a diameter of 24 c. m. and 22½ c. m. above, it is 20¾ under the 4th. There is, therefore, a slow and gradual diminution of the diametral size of the articulations, from the lower to the upper one which is more abruptly inflated, and a gradual diminution of the stem in its whole length from 23 c. m. at base to 20 c. m. at the upper part. The stem is somewhat compressed or oval, being 9 c. m. in diameter in one direction, and 4½ in the other. The scars, which are in spiral order, as in all the species of the group, are less distinct and their direction is changed and irregular at the articulations, though quite regular in the internodes.

This peculiar inflation of the stem at irregular intervals and the irregularity of the leaf scars upon the knots can result only from an annual growth which, becoming gradually less active upward, as seen from the diminution of the diameter above the articulations, is increased in diameter upon the knots where the vegetation has stopped during a certain lenth of time, to begin again under more favorable circumstances, in the same way as we see it upon the young trunks or the branches of pines. The same process is remarked upon other stems of *Sigillaria*, like that of *S. mamillaris*, described "Coal Flora," p. 484, Pl. LXXII, Fig. 5, where the period of rest or interruption in the vegetation is indicated by the presence of adventive, undeveloped buds of leaves.

Habitat—The location of the specimen is indicated by Mr. Huston as being about 50 feet below the Crinoidal limestone of Ohio.

SIGILLARIA MAMILLARIS, *Brgt. Pl. CVIII, Fig. 6.*

Lesqx. "Coal Flora," p. 483, Pl. LXXII, Figs. 5, 6.

The specimen figured is in a perfect state of preservation and clearly shows the variability of the species, first in the more or less distant space between the areoles and then in the gradual modification of their shape. The figured fragment is 14 c. m. long. The costæ are plano-convex, 6 to 7 m. m. broad, with narrow undulate furrows; the areoles at the base of the fragment are 5 m. m. distant, while in the upper part the space between them is not more than 1 m. m. Toward the base the areoles are oval, truncate above, rounded downwards and very obtuse at base, 6 m. m. long, 3 m. m. broad in the lower part, while becoming gradually shorter and more distinctly angular, the upper ones are regularly hexagonal, 3 to 3½ m. m. broad and as long. The corticated surface preserves the same characters in all the forms of this species.

Habitat—Large specimens have been procured by Superintendent Roy from the coal mines of Butchel, Ohio, for the cabinet of Mr. R. D. Lacoe.

SIGILLARIA PYRIFORMIS, *Brgt.*

Stem costate; costæ 8–10 m. m. broad; furrows obtuse, sinuous, distinct; cortex striate lengthwise, traversed by an arcuate transversal line above the scars, and by two vertical lines or sometimes transverse wrinkles under them; areoles not distant, discoid, oblong, obtuse, enlarged in the lower part, subpyriform; vascular scars 3, placed in the upper part of the areoles, the middle small, punctiform, the lateral arcuate.

The specimen representing this species has the costæ 8 m. m. broad, the furrows slightly undulate, the areoles ovate, pyriform, slightly emarginate at top, 9 m. m. long, 6 m. m. broad toward the base, where the lower border forms a short narrow ridge passing obliquely from base of the areoles to the furrows. The vascular scars are placed quite near the upper border; the distance between them is equal to their length; their form is somewhat like that of *S. Lescurii.*, "Coal Flora," Pl. LXXII, Fig. 8, but there is a marked difference in the narrower sinuous costæ, the comparatively longer and narrower areoles only slightly enlarged below the middle.

Habitat—The specimen described is No. 1312 of the National Museum. Its locality is not indicated. Another specimen less distinct, No. 901 of Lacoe's collection, from Plymouth, Penna., F vein, also represents the species.

SIGILLARIA LEVERETTII, *Sp. nov., Pl. CVIII, Figs. 4–5.*

Stem costate; costæ sinuous, transversely rugose, 2 c. m. broad; separated in the young stems by narrow deep furrows 1 or 2 m. m., in old stems, distantly separated by a very coarsely rugose bark, covering a flat irregularly striate surface, sometimes passing above the ribs and covering their areoles which are 2 to 2½ c. m. distant vertically. These are broadly rhomboidal in outline, acute at the apex and the angles, truncate or obtuse at base, with three vascular scars, the central punctiform, traversed by a horizontal line, the lateral semi-lunar.

This beautiful species, without analogy to any other described, shows two different aspects in the specimens figured, differences evidently resulting from the position of the fragments of bark upon the trunk. In Fig. 4, the costæ are distant, separated by a layer of thick bark very coarsely striated and undulate lengthwise, and the areoles, of which the lower part is obliterated, are broadly, regularly rhomboidal, the upper lines, joined in an obtuse angle at the top passing down to the lower in rounding at the sides. The vascular scars are indicated by two large lateral semilunar appendages inflated at base, joined at the apex by a deep horizontal line; the base of the areoles is truncate and joined to a semi-globular appendage. In Fig. 5, the areoles are more distinctly preserved and formed of an upper rhomboidal scar of which the upper lines are joined in the middle into a short acute point, the lower abruptly turning toward the vascular scars, while the areoles are prolonged downward into a broad obtuse base, enclosing the vascular scars in the middle. The ribs between the areoles are more or less deeply transversely rugose, the wrinkles being thin and close, interspersed with some more distant, thick ones.

Habitat—Des Moines county, Iowa, communicated by Frank Leverett in good specimens now in the collection of Mr. R. D. Lacoe, No. 480

SIGILLARIA WILLIAMSII, *Lesqx., Pl. CVII, Fig. 15.*

"Coal Flora," p. 488.

To the description of the species, l. c., I have merely to add: that the cortical surface is regularly striate and that, with a magnifier, the striæ are indistinctly perceived through the concave surface of the ribs. The furrows are about 2 m. m. deep, 1 m. m. wide at the concave bottom; the leaf-scars are not as angularly notched as marked upon the figure; the subcortical scars are obscurely oval, centrally marked by a round or oblong mamilla.

DIDYMOPHYLLUM (SIGILLARIA) OWENII, *Lesqx., Pl. XCII, Fig. 11.*

"Coal Flora," p. 507, Pl. LXXIV, Figs. 10-10b.

The trunk described under this name is here represented exactly with the measurements of its different parts. The circumference of the trunk, near the base, is 1, 12 m.; the preserved standing portion 70 c. m. high, above the roots which abruptly curve to an horizontal direction in 6 or 7 primary branches 12 to 24 c. m. in diameter. They are first forked at 20 to 30 c. m. from their point of divergence from the trunk and measure at their ends, where they are broken, 3 to 8 c. m. in diameter. The representation of the roots in "Coal Flora," l. c., was not made from exact measurements; these procured for me a long time ago (1861) by the kindness of Mr. John Chappells Smith, of New Harmony, having been for a long time missing. As the specimen has been lost in the destruction by fire of the Museum of the Indiana University, that copy is worth to be preserved to Paleontology. The leaf-scars are represented first in a reduced measure, proportionally to that of the stems. Fig. 1, 1*a*, show them, natural size; Fig. *b*, enlarged; Fig. *c*, represents the scars nearly at the origin of the roots; Fig. *d*, the scars upon the roots.

CORDAITEÆ.

"*Coal Flora*," p. 525.

CORDAITES, *Ung.*

"*Coal Flora*," p. 527.

To what has been already published on the plants of this group, in "Coal Flora," l. c., I can add only the descriptions of a few species of fruits and flowers recently found and as yet unknown.

CORDAIANTHUS, *Grd' Ey.*

Lesqx. "*Coal Flora*," p. 544.

CORDAIANTHUS FLEXUOSUS, *Sp. nov.*, *Pl. CIX, Fig. 2.*

Raceme simple, long and slender, flexuous; nutlets or flowers round, oval, 5 m. m. long, closely involucrate by oblong obtuse scales apparently attached to the base of the nutlets in a single row, not longer than the nutlets and covering them like a sepaloid involucre.

The raceme is about 17 c. m. long; the buds or ovules opposite, 1 c. m. distant, gradually a little nearer to each other toward the apex, the rigid bracts being at right angles to the narrow rachis.

Habitat—Campbell's ledge, near Pittston. Cabinet of Mr. R. D. Lacoe, No. 846.

CORDAIANTHUS SPICATUS, *Sp. nov.*, *Pl. CIX, Fig. 1.*

Involucrate nutlets, distichous, or subalternately placed on both sides of a thick rigid thinly striate stem or rachis, in the axis of linear involucral leaves nearly as long as the nutlets and indistinct.

The seeds are 1 c. m. long, not half as broad, ovate, acute, convex and somewhat carinate in the middle; the involucres are composed of imbricate narrowly lanceolate or acuminate scales apparently embracing the ovules from the base and covering them; the stem is 5 m. m. broad, the ovules 8 m. m. distant, forming a fragment of spike 15 m. m. long.

This species resembles *Botryoconus femina*, Grd'Ey.,

"Fl. Carbon," p. 279, Pl. XXXIII, Fig. 1, so much, indeed, that at first sight I was disposed to consider it as identical. But the species of Grand'Eury represents small strobiles, formed of imbricated concave scales, covering each a very small seed axillary placed on their lower side, while the involucrate seeds of this species are true ovules, with a smooth very convex surface, not flattened by compression as would be a strobile composed of scales. The name of *Botryoconus* is therefore not applicable to this kind of fructification, that genus being proposed by Goeppert for branches bearing spikes composed of imbricated scales, like the *Antholithes* of authors. The seeds of this species are smaller than those which have generally been described as *Cordaicarpus*, but not smaller than those figured by Grd' Ey. as *C. congruens* "Fl. Carb., Pl. XXVI, Fig. 21 ; seeds which like those of Fig. 1, l. c., are somewhat carinate on the back. The relation is also marked with the fragment of a spike in Newberry, "Geol. Rep. of Ohio, Paleont." 1, Pl XLI, Fig. 2, described p. 41 as *Antholithes*. The seeds of this last spike are oval, hard nutlets, enclosed at base in large short scales, in the same manner as in our Fig. 1, the difference being merely in the size and shape of the nutlets and of the scales.

This and the following species are referable to the group *Cordaianthus baccifer*, Grd.'Ey.

Habitat—Campbell's Ledge, Pittston, No. 940 of Lacoe's collection.

CORDAIANTHUS RUGOSUS, *Sp. nov.*

Stem large, distinctly transversely rugose ; nutlets close, subopposite, distichous, oval or oblong, obtuse, short pedicellate, axillary and involucrate ; lower bracts linear-lanceolate, twice as long as the ovules, involucre indistinct, apparently composed of basilar imbricate short lanceolate scales.

The ovules, as seen separated upon the stone and free of their involucre, are 4 m. m. long and two broad, the pedicel is short, ½ m. m. The species is also somewhat like the *Antholithes* of Newby., l. c., Pl. XLI, Fig. 2, already mentioned above, differing by the ovules longer and narrower, the indistinct involucre seemingly absent and the stem transversely rugose.

To this species I refer as a variety a stem of the same character as the preceding but lineate lengthwise, and rough-punctulate, the points looking like scars of short hairs, the ovules being slightly longer and the involucre also indistinct.

Habitat—The first specimen No. 929 of Lacoe's cabinet is from Shamokin, Pa; the second, No. 943 of the same collection, is upon a nodule of Mazon creek.

CORDAICARPUS, *Grd' Ey.*

Lesqx., "Coal Flora," p. 549.

CORDAICARPUS GUTBIERI, *Grd' Ey., Pl. CIX, Fig. 3.*

Lesqx., "Coal Flora," p. 549, Pl. LXXXIII, Figs. 8-11.

The specimen represented Fig. 3 is the best preserved one found among a very large number of seeds of the same kind which have been abundantly collected at Cannelton. The fruits are generally obliterated in their form by compression.

CORDAICARPUS COSTATUS, *Lesqx., Pl. CIX, Fig. 4.*

Cordaites costatus, Lesqx., "Coal Flora," p. 540, Pl. LXXXVI, Figs. 1 and 2.

The seed Fig. 4 is in a better state of preservation than that published attached to the stem l. c. The central nucleus is exactly oval, and the surface convex, surrounded by an inflated, sometimes half round, border. The specific name is that of the *Cordaites* to which it was found attached, whose stem is remarkable by the prominent bolsters, impressions of the base of fruits and leaves by which old stems have become marked by distinct costæ. The fruit is figured overturned.

Habitat—Very common at Cannelton, also found at Lorberry Junction, Schuylkill county, and at Everhart Creek, near Pittston, Pennsylvania. No. 984 of Lacoe's collection.

CORDAICARPUS CINCTUS, *Sp. nov., Pl. CIX, Figs, 5, 6*

Seeds exactly round, quite flat, attached to a slender, flexuous narrow rachis, bordered all around by a narrow rim; surface smooth, rather polished.

The stem $2\frac{1}{2}$ m. m. broad, bears 2 pairs of seeds $1\frac{1}{2}$ c. m. in diameter, exactly circular, attached by a slightly inflated narrow border and subopposite. A separate seed, Fig. 6, has the same character. The broken part, seen at its upper border, seems to have been lacerated at the point of attachment.

The seed is referable to the genus *Cyclocarpus*, Gœpp., which has been united by Grd 'Ey. to *Cordaicarpus* and by Schimper to *Cardiocarpus*. All the seeds of a character similar to the one described above are apparently fructifications of *Cordaites*. Even many specimens of *Cordaicarpus Gutbieri* are nearly round seeds, as seen, "Coal Flora," Pl. LXXXIII, Figs. 10–11.

Habitat—Found in digging down a bed of shale for a cellar; horizon of Coal E, Pittston, Pennsylvania. The specimen Fig. 1 is No. 939 of Lacoe's cabinet.

CORDAICARPUS STABILIS, *Sp. nov.*

Seeds smaller than those of the preceding species, 8 m. m. in diameter, with exactly the same round form, the border narrower, less distinct, the surface smooth.

These seeds are very numerous in the shale of the Arkansas sub-carboniferous coal, mixed with *C. ovatus*. All the specimens show the same form and the same size.

Habitat—Male's coal, Arkansas, F. L. Harvey.

CORDAICARPUS LINEATUS, *Sp. nov., Pl. CXI, Fig. 16.*

Nucleus exactly cordiform thinly striate lengthwise.

This seed, apparently a detached nucleus, seems referable to this genus. It is $1\frac{1}{2}$ c. m. long 12 m. m. broad in the widest part toward the base. It is comparable to the nucleus of the seed of *Jordania bignonioides*, Fiedler, "Foss," Fruits, Pl. XXVIII, Figs. 37, 44, but still more to Grd'Ey., *Cordaicarpus major*, "Fl. Carb.," p. 235, Pl. XXVI, Fig. 16, from which it differs merely by the more deeply cordate and more slender form of the seed with its surface thinly striate.

Habitat—Cannelton, Pa., No. 893 in Lacoe's collection.

CARDIOCARPUS, *Brgt.*

Lesqx., "Coal Flora," p. 561.

The definition given in "Coal Flora" should be modified as follows:

Seeds of various shape, composed of a compressed generally cordiform or oval, acute or acuminate nucleus surrounded by a flattened fibrous border or a membranaceous wing and often narrowed at base into a short pedicel or acute appendage.

This modification is forced by the discovery of a number of seeds appearing pedicellate at the base, by the elongation and contraction of the wing, as in Pl. CIX, Figs. 13 to 15. In his description of the genus, Brongniart says of the nucleus, that it is short-pedicellate at the emarginate base. The prolongation of the nucleus into the narrowing part of the wing may be observable by anatomical section of silicified seeds; but I have only seen it upon Figs. 13–17 of the same plate and always narrowed, acuminate, ending into the border of the wing or above its base. In some other seeds, as in Figs. 9, 11, 15, this so-called pedicel is represented by a mere round small tubercle and the wing is not prolonged downward; while still in others, the wing is prolonged into what seems to be a pedicel, and the base of the nucleus remains rounded or emarginate as in many of the figures of Pl. CX. These different variations which are not often perceivable upon compressed specimens have forced me to preserve in this genus a number of seeds described as *Samaropsis* by authors, or under different names by Brongniart, these last from the characters of the inner structure.

CARDIOCARPUS DILATATUS, *Sp. nov., Pl. CX. Fig. 2.*

Nucleus comparatively small, broadly ovate or subreniform, sharply pointed, bordered all around by a narrow, convex, and continuous wing, enlarged on the sides, comparatively very broad, emarginate at the base, rounded and narrowed in curving to the point of the nucleus.

The nucleus is 10 m. m. vertically, 15 m. m. in horizontal diameter, without the border which is 1 m. m. broad. As seen from the cavity left around the nucleus, the specimen being an impression, the border was thick or highly convex. The wing is in the middle, 1 c. m. in diameter, narrowed to 2 m. m. at the emarginate base and also at the apex, where it curves abruptly to the top of the nucleus, the point of junction of the wing being there also 2 m. m. broad. This beautiful seed is remarkably similar to *C. Baileyi*, Daws., "Geol. Survey of Canada, fossil plants," p. 60, Pl. XIX, Fig. 219, from which it differs by the much larger nucleus with a distinct narrow rim separating it from the wing.

Habitat—Sub-carboniferous measures. Arkansas and Campbell's Ledge, near Pittston, Pa. No. 961 in Lacoe's collection.

CARDIOCARPUS PATENS, *Sp. nov., Pl. CX, Fig. 3.*

Nucleus nearly round or emarginate at base, there marked with a short round tubercle; wing narrow, converging and narrowed at the base, which is destroyed, broadly emarginate at top to a little above the border of the nucleus.

The central seed is 14 m. m. in diameter, round, not acuminate nor acute at the apex; the wing broken at base is 3 m. m. wide, joined at the apex by 1 m. m. only of its diameter and traversed by a thick line slightly enlarged on the lower part, passing up from the middle of the nucleus (the nucelle), to the micropile. The inner part of the nucleus is marked by two arched lines diverging from above the nucelle and parallel to the borders, forming like a double testa, as seen in the seed described by Brongniart, Graines foss., "Ann. des Sci. nat. 5e Ser. Bot.," Vol. 20, p. 14, Pl. 21, Figs. 12–13, as *Diplotesta.*

Habitat—Arkansas Coal measures. F. L. Harvey.

CARDIOCARPUS SPECIOSUS, *Sp. nov., Pl. CX, Fig. 1.*

Ovule large, ovate, acute, deeply cordate or emarginate at base, surrounded by a broad margin or wing, parallel to the borders and continuous.

The nucleus is 5 c. m. long, 3 broad in the middle; the wing, flat and smooth, is 1 c. m. in the middle, narrowed to 5 m. m. at the apex. It is partly destroyed on one side and at base, but appears to have been continued, as it is larger at the base than in the middle of the seed.

I consider this seed to be closely related to that described "Coal Flora," p. 593, Pl. LXXXV, Fig. 16, under the name of *Carpolithes bifidus*, Lesqx., if not the same species. The name of *C. bifidus*, as originally described "Geol. of Penna.," p. 877, Pl. 17, Fig. 10, cannot be preserved, as it was given to a very poor specimen of an undeveloped or crushed fruit, apparently a nucleus, split at the apex by compression and whose relation is very uncertain.

Habitat—Cannelton, Pa. I. F. Mansfield.

CARDIOCARPUS HARVEYI, *Sp. nov., Pl. CIX, Figs. 22, 23.*

Fruit large, oval in outline; nucleus oval, slightly emarginate at base, narrowed at the apex and bordering the micropyle; wing flat, broad, deeply emarginate at top into two pointed erect horn-like divisions.

The nucleus of the best preserved of these fruits is 2½ c. m. broad, 4½ c. m. long to the base of the notch. The borders are 9 to 10 m. m. broad, quite flat, smooth, while the central part or the nucleus is irregularly verucose on the surface. Fig. 23 differs by the nucleus narrower, obtuse at both ends and the wing not emarginate at base. The difference is probably caused by the degree of maceration and the angle of compression.

Habitat—Arkansas Coal measures, sub-conglomerate. Prof. F. L. Harvey, Tracy, Tennessee. No. 299 of Lacoe's collection.

CARDIOCARPUS LONGICOLLIS, *Sp. nov., Pl. CIX, Figs. 24, 25.*

Seeds large, bottle-shaped; nut large, oval or ovate, rounded or truncate at base, smooth or rugose on the surface, prolonged above into a long neck or micropyle; margin broad, continuous and enlarged at base, ascending upward along the neck and bordering it.

The nucleus is 22 m. m. broad, 4 c. m. long to the base of the neck; the wing, 5 m. m. broad, is quite flat and smooth, somewhat enlarged at the base (broken), narrower along the neck, where it measures only 1½ to 2 m. m. This seed does not belong to this genus, but its affinity is unknown to me and I describe it here on account of the broad margin bordering the nucleus, by which it is at least related to the last species. A longitudinal section of a seed, on which Brongniart has established his genus *Tripterospermum*, "Gr. foss," p. 19, Pl. 22, Fig. 6, is much like the seed described above. But *Tripterospermum*, says the author, has the general form of a *Trigonocarpus* and the nucleus without the testa would have the same character; but the testa is very thick and prolonged into three projecting wings. Nothing on these seeds indicates such a kind of division; the projecting wings or ribs like those of *Trigonocarpus* being generally more or less observable, even upon flattened seeds, as, for example, those of Pl. CIX, Figs. 19–21.

Habitat—Sub-conglomerate measures of Tennessee. No. 798[a] and 798[b] of Lacoe's collection.

CARDIOCARPUS PACHYTESTA, *Lesqx.*, *Pl. CIX, Figs. 13–15.*

"Coal Flora," p. 565.

The description in "Flora" is given from less perfect specimens. The nucleus is rarely round, more generally ovate, even narrowly ovate-acuminate, and from its base it is continued downward into a narrowly lanceolate acuminate appendage or axis of a pedicel, the borders or wing being also prolonged downward to the point of attachment.

Habitat—Campbell's Ledge, near Pittston, Sub-conglomerate. No. 293 of Mr. Lacoe's cabinet.

CARDIOCARPUS ELONGATUS, *Newby.*, *Pl. CIX, Figs 16–17.*

"Coal Flora," p. 567, Pl. LXXXV, Fig. 41.

As for the preceding species, better specimens have been obtained since the first description. These show, Fig. 16 the basilar appendage of the nucleus, the prolongation of the border being destroyed, and. Fig. 17 both the appendage and the pedicel. In this last figure the top of the wing is slightly notched; but I do not consider that as a specific character. The same difference is observed in a number of specimens of a closely allied species, *Cardiocarpus cornutus*, Daws., "Geol. Survey of Canada, fossil plants," Pl. XIX, Figs. 14, 15, some of which have the border joined at the apex, while others are deeply emarginate and acuminate, the divisions being either diverging or connivent at apex.

Habitat—Campbell's Ledge, Pittston. No. 271 of Lacoe's collection.

CARDIOCARPUS MAMILLATUS, ? *Lesqx., Pl. CIX, Fig. 7.*

"Coal Flora," p. 571, Pl. LXXXV, Figs. 32–33a.

From a number of well preserved specimens obtained at different localities, it has become evident that Figs. 32 and 33 of "Coal Flora," l. c., do not represent the same species, as I supposed it, Fig. 32 being a *Cardiocarpus*, seen with its testa or border partly preserved, as in Pl. CIX, Fig. 7, while Fig. 33 is referable to *Rhabdocarpus*, this being always found without any border or testa.

Habitat—The specimen figured, Pl. CIX, was sent from Arkansas subconglomerate coal by Prof. F. L. Harvey, a different stage of the Carboniferous from that where *Rhabdocarpus mamillatus* has been obtained.

CARDIOCARPUS OVALIS, *Sp. nov., Pl. CIX, Figs. 8, 9.*

Seeds small, nucleus oval or ovate, blunt at apex or obtuse; border continuous, a little narrower at base, emarginate at apex into a minute semilunar notch.

The nucleus Fig. 8 is 10 to 12 m. m. long, 8 m. broad below the middle, truncate or slightly emarginate at base; the ring, Fig. 9, is 1½ m. m. broad towards the base, 2½ m. m. at the apex. The point of attachment of the pedicel or appendage is marked by a round depression at the base of the nucleus. The species is allied to the following.

Habitat—Coal measures of Arkansas. F. L. Harvey.

CARDIOCARPUS CONGLOBATUS, *Sp. nov., Pl. CIX, Figs. 10, 11.*

Differs from the preceding merely by the exactly round shape of the nucleus and the borders nearly of equal width all around. The apicial notch is not semilunar but rectangular, also very small. In both these species the nucleus is often found separately; its surface is somewhat convex.

Habitat—Same as the preceding species.

CARDIOCARPUS DIVERGENS, *Sp. nov., Pl. CX, Fig. 4.*

Nucleus broadly ovate, acutely pointed; border parallel to the nut, equal in width or slightly larger toward the apex, joined at the top of the nucleus, and there emarginate, the borders diverging at right angles.

The fruit is about the same size and form as that of *C. patens*, Fig. 3. But it is of a different structure, the nut being ovate, acute, without trace of nucelle or of micropile and not pedicellate nor tuberculate at base. The nut is 1 c. m. long, 12 m. m. broad; the border 3 m. m. wide.

The species is related to *Cardiocarpus* (*Cyclocarpus*) species, figured in Schp. "Paleont. Veget. Atl," Pl. CX, Fig. 21, which, however, is at least twice as large.

Habitat—Pittston, E vein and Boston mine C vein. No. 964 of Lacoe's collection.

CARDIOCARPUS LATIOR, *Sp. nov., Pl. CX, Figs. 5, 11, 12.*

Seeds broadly ovate; nucleus round or ovate, obtuse or with a short acute appendage at base; borders narrow at and toward the base, gradually enlarged upward to the apex where they are connate, except at the upper rounded border.

Fig. 5 is a concave impression showing the outer testa from which the nucleus has been separated. In Figs. 11 and 12 the nucleus is still present, and in Fig. 12 the fruit is partly covered by fragments of the outer testa, transformed into a comparatively thick pellicle of coal. In Fig. 12 the nucleus is prolonged at base into a short-pointed appendage. These three fruits may represent different species, the nucleus being longer and narrower in Fig. 5, shorter, broader, and nearly round in Figs. 11 and 12.

Habitat—Sub-conglomerate coal measures of Tennessee. Lacoe's No. 977.

CARDIOCARPUS CRASSUS, *Sp. nov., Pl. CIX, Fig. 12, CX, Figs 6–9.*

Small seeds. Nucleus ovate, acutely pointed, round or truncate at base; borders comparatively large, rarely distinct, continuous, of the same width all around, emarginate at the apex and separated only by the narrow micropyle marked by a line often continued downward to the base.

The nucleus is 5 to 8 m. m. long, 5 to 6 m. m. broad, very variable in form, sometimes surrounded by a ring or inner testa, as in Fig. 8. The borders are more or less emarginate, some at right angles from the base, as in Fig. 12; others connate to the apex, as in Fig. 9. They have been found either separated or spread upon a large piece of shale where all these forms and other intermediate ones have been observed. They resemble *C. simplex*, "Coal Flora," Pl. LXXXV, Figs. 48–50, differing essentially by the thick inflated or convex surface of the nucleus, and the narrow border less deeply or broadly emarginate. It is, however, difficult to separate species of *Samaropsis* and perhaps *Cardiocarpus zonulatus*, *C. late-alatus*, of the "Coal Flora," and *C. crassus*, may represent one species only, though different they may appear. The nucleus of this species is often found separated from its testa.

Habitat—Campbell's Ledge, Pittston. No. 963 of Lacoe's cabinet.

CARDIOCARPUS CIRCULARIS, *Sp. nov., Pl. CX, Fig. 10.*

Nucleus nearly round, minutely pointed, 6 m. m. in diameter; ring 2 m. m. broad, equal and continuous to the top where it is truncate above the point of the nucleus.

The seed resembles a large form of the preceding species; it is, however too, large and comes from a different locality at a different stage of the coal measures.

Habitat—Vermillion county, Indiana. No. 976 of Lacoe's cabinet.

CARDIOCARPUS DIPLOTESTA, *Sp. nov., Pl. CX, Fig. 13.*

Nucleus small, round; inner testa thick, acuminate by a micropylar prolongation passing up between the lips of the bordering testa which is very narrow at base, enlarged upward, connate and very slightly emarginate at apex.

This species is constituted like the following which it resembles, being essentially different by its larger size. The nucleus, placed lower than the center, is 5 m. m. in diameter, perfectly round; the inner testa narrowed at base, where it is a little more than 1 m. m. broad, is gradually enlarged upward to 3 m. m., at the point of connection where it is traversed as well as the borders, by a narrow micropyle ascending to the apex. The borders, representing an outer testa, are very narrow, scarcely 1 m. m. in the lower part of the seed, widened in the upper part to 3 m. m., and there connate in nearly their whole width. I have seen only one specimen of this species.

Habitat—Campbell's Ledge, Pittston. Collection of Lacoe, No. 272.

CARDIOCARPUS ZONULATUS, *Lesqx., Pl. CX, Figs. 14–17.*

"Coal Flora," p. 568, Pl. LXXXV, Figs. 44, 45.

The specimens figured are more complete than those which I had for the description of the species, l. c., and which all had the base destroyed The nucleus is oval or broadly cordiform, acute; the inner testa is distinct, more or less narrow, the outer testa or ring comparatively broad, larger in the upper part and toward the base, where it passes into a short pedicel. The nucleus is often marked in the middle by a vertical line which rarely reaches the apex of the seeds; more generally still by a central depression in the place of the nucelle.

Habitat—Same as the preceding, represented by numerous specimens in the collection of Mr. Lacoe, No. 272.

CARDIOCARPUS BICUSPIDATUS, *Sternb., Pl. CX, Figs. 18–22.*

Lesqx., "Coal Flora," p. 573, Pl. LXXXV, Figs. 42–43.

From a very large number of well preserved specimens, the characters have been clearly seen. The nucleus is from 1 to nearly 1½ c. m., transversely measured; from 7 to 15 m. m. vertically. It is often separated from its testa, and very smooth. The border of the testa is narrow, rarely broader than 1 m. m., except in the widening base or pedicel.

Habitat—The species has been found in abundance in the shale of the coal of Arkansas, near Fayetteville, by F. L. Harvey. No. 985 of Lacoe's collection.

CARDIOCARPUS ELLIPTICUS, *Pl. CX, Figs. 23–27.*

Carpolithes ellipticus, St. "Fl. d. Vorw." 1, p. 40, Pl. VII, Fig. 1.

Fruit small; nucleus elliptical obtuse at both ends; border narrow, equal, slightly emarginate at one end.

The nucleus is 6 to 8 m. m. long, 4–5 m. m. broad in the middle; the flat border, about 1 m. m., is a little narrower on one side. Figs. 23 and 24 represent the normal form with the surface of the nucleus smooth. Figs. 25 and 27 have the surface rugose, the nucleus somewhat broader, the border very narrow, even invisible toward the upper part. These last specimens may represent a different species.

Habitat—Cannelton, Penna., No. 968; the var. 994 is from Campbell's Ledge, of Pittston, both of the collection of Mr. Lacoe.

CARDIOCARPUS ANNULARIS, *Pl. CX, Figs. 28–30.*

Carpolithes annularis, Sternb., "Fl. d., Vorw.," 1, p. 4 Pl. VII, Fig. 15.

Seeds a little larger than those of the preceding; nucleus oval, with an equal continuous border, sometimes prolonged downward into a short pedicel; surface smooth or rugose.

These fruits are much like the last described, but larger, 1 c. m. long, 8 m. m. broad, including the border which is nearly 2 m. m. broad. The seed Fig. 29 appears different, but it was found mixed with a number of those of the normal form, some among them intermediate in shape. The lower prolongation of the border or pedicel is rarely preserved in the species of this genus. The figure of Sternberg, l. c., bears at its base an oval tubercle, about like the appendage of Fig. 29.

Habitat—Campbell's Ledge, Pittston, sub-conglomerate. No. 986 of Lacoe's collection.

CARDIOCARPUS PUSILLUS, *Sp. nov., Pl. CX, Figs. 31–33.*

Seeds very small, 2–3 m. m. in diameter, broadly oval or nearly round; nucleus globular and central, 1½ to 2 m. m., bordered by a distinct narrow ring.

The central part or nucleus is convex, semi-globular in appearance, sometimes oval or more elongated; the narrow ring is continuous around the nucleus and parallel to the borders, equal. The seeds are coriaceous, shining.

Habitat—Not rare at Cannelton, Pennsylvania. Lacoe's collection, No. 967.

RHABDOCARPUS, *Goepp. and Berg.*

Lesqx., "Coal Flora," p. 574.

RHABDOCARPUS LATE-COSTATUS, *Sp. nov., Pl. CX, Figs. 34, 35.*

Seeds small, oval, broadly costate, mamillate at the apex, surrounded by a narrow flat border.

The fruit is only 7 m. m. long, 4 m. m. broad in the middle, marked upon its exposed surface with three comparatively broad costæ, tipped with a small projecting mamilla, or an areole after abrasion.

Habitat—Campbell's Ledge, sub-conglomerate. No. 969 of Lacoe's collection.

RHABDOCARPUS INFLATUS, *Sp. nov., Pl. CX, Fig. 36.*

Fruit large, ovoid, inflated, and broadly round at one end, obtuse at the other, transversely wrinkled, as a small bladder would be by compression, obscurely regularly lineate lengthwise.

The fruit still bears its outer testa which appears to have been soft or somewhat fleshy, being transversely folded or undulate by compression. The shape is somewhat like that of *Rhabdocarpus Mansfieldi*; but the characters, especially the size and the obscure striæ, are different.

Habitat—Cannelton. No. 991 of Mr. Lacoe's cabinet.

RHABDOCARPUS PACHYTESTA, *Sp. nov., Pl. CX, Figs. 37, 38.*

Seeds large, ovate, with a thick testa marginate at apex, rounded at base, enclosing an oval nucleus which is regularly striate lengthwise, alveolate at apex.

The seed, 3½ c. m. long, 2 c. m. broad, is in a ferruginous concretion and shows the structure as represented by the figure. The nucleus is only 13 m. m. long, 9 m. m. broad, loosely attached to the outer testa, which appears as split into two halves, the two borders being flat while the nucleus remains entire and prominent in the middle. The testa is 9–10 m. m. thick, above and below the nucleus, 6 m. m. on its sides. The nucleus, easily detached, much resembles the seeds of the following species and may represent it differing, however, not only by the testa, which though never found with the seeds of *R. mamillatus*, might be supposed to have been destroyed by maceration, but essentially by the deeper more distant striæ of the inner nut. Fig. 38 shows the fruit as seen obliquely.

The preservation of the outer testa in its whole thickness, as it is in this seed fossilized into pyrite, allows us to understand how far different are the characters of the fructifications of the Carboniferous, as seen in their original state, from what they appear to be when they have been softened by maceration and flattened by compression. That will also give an idea of the prodigious reduction in bulk of the soft cellular tissue by the process of carbonization or of transformation into coal. In this case, if the shrinking should be equal to that indicated by Stur for the carbonization of the bark of *Calamites*, that outer testa being transformed into coal would be represented by a pellicle of coal less than one fourth of a m. m. in thickness.

Habitat—Nodules of Mazon creek, No. 989 of Lacoe's collection.

RHABDOCARPUS MAMILLATUS, *Lesqx., Pl. CX, Figs. 39–42.*

Cardiocarpus? mamillatus, Lesqx.," Coal Flora," p. 571.

In the description of the species, "Coal Flora" l. c., I have confounded two apparently different forms; one a *Cardiocarpus* described above, the other a true *Rhabdocarpus.*

At least a number of specimens have been found all representing the species as it is now figured, and as it has been originally described in "Geol. Rep. of Ills.," IV., p. 461, Pl. XXXI, Figs. 12–15. The specimens are embedded into a generally loose soft-grained sandstone rock, and covered by a thin amorphous coating of brown matter under which appears the inner testa generally absent, but which, when preserved, is a thin layer of coal, smooth on the surface, covering the amorphous sandstone composing the nucleus. This is oval, a little more enlarged and obtuse at one end, marked at the larger end by a round areole, like a point of attachment, more acute and mamillate on the other. The nucleus is equally regularly striate, its size varies from 10 to 15 m. m. in length and from 7 to 10 m. m. in width.

Habitat—Stark county, Illinois, No. 665 of Lacoe's collection.

RHABDOCARPUS SUB-GLOBOSUS, *Sp. nov., Pl. CX, Figs. 43–45, 62.*

Seeds of medium size broadly ovoid, more enlarged in the middle, obtuse or rounded at both ends, equally but obscurely lineate lengthwise, the striæ at equal distance.

The seeds are sometimes covered with a thin pericarp, reduced to a thin pellicle of coal, as in Fig. 43, and as seen in that figure also, the striæ, generally 1 m. m. distant, appear sometimes separated by thin close intermediate lines. The seeds vary from 10 to 13 m. m. long and from 7 to 9 broad. One of the specimens, Fig. 62, which I refer to this species with some doubt is alveolate at the apex ; but it is so much like Fig. 45, which also appears to be truncate or alveolate at top, that a separation does not seem legitimate.

The species is comparable to *R. ovoideus*, Goepp. and Berg. "De fruct.," p. 22, Pl. I, Fig. 17, and Goepp. "Perm. Fl.," p. 173, Pl. XXVII, Figs. 9, 10, from which it differs by the size of the seeds which in the American species are twice as large, and by the distinctly lineate surface.

Habitat—Cannelton, Pa., and Butler Dam C. vein of Pittston. No. 978 of Lacoe's collection.

RHABDOCARPUS EMARGINATUS, *Sp. nov., Pl. CX. Figs. 46, 47.*

Seeds oval oblong, bordered by a tumid margin somewhat distinct only in the upper part, emarginate at the apex, closely striate lengthwise with thin parallel lines; outer testa marked by thin flakes of coaly matter, striate on the surface like the nucleus.

The two specimens figured are the only ones seen of this species. They are 2½ c. m. long, 10–12 m. m. broad in the middle. The slightly inflated border, 2 m. m. in diameter, is distinct, in Fig. 47, all around the seed, except at base. In Fig. 46, it is seen merely in the middle and there obscurely. As it is striate exactly in the same way as the inner part, it does not appear to constitute a separate ring as in species of *Cardiocarpus.*

Habitat—Cannelton, Pa. No. 233 of Mr. Lacoe's collection.

RHABDOCARPUS TENAX, *Sp. nov., Pl. CXI, Figs. 25, 26.*

Resembles the last species, differing by its irregular form, oblong or oval, truncate at base, obtuse or sub-emarginate at the apex, its size smaller, and the striæ less distinct and more unequal, some of them, especially those in the middle of the seeds, being deeper and broader. The seeds 16 m. m. long, half as broad, are generally found in great numbers and placed as if depending from a thick rachis generally destroyed. The only fragment preserved of it is represented Fig. 25. It is evidently part of a large rachis upon which the sessile seeds are attached about like those of Pl. CIX, Fig. 1.

Habitat—I have large specimens of this species from Cannelton, communicated by Mr. I. F. Mansfield. It is also in Mr. Lacoe's cabinet, No. 701.

RHABDOCARPUS ABNORMALIS, *Sp. nov., Pl. CX, Figs. 48, 49.*

Fruit large, of irregular shape, oblong, abruptly narrowed and obtuse at one end, somewhat curved in the lower part, obtusely pointed at the base, thinly, equally lineate lengthwise, and partly covered by a thin coaly layer 1½ m. m. thick, remains of a pericarp.

The two seeds figured are upon the same piece of shale. They are 6 c. m. long, 2½ broad in the upper part, resembling large teeth or bones, but evidently plant-remains or large fruits of unknown affinity.

Habitat—Cannelton, Pa., Mr. I. F. Mansfield.

RHABDOCARPUS APICULATUS, *Sp. nov., Pl. CX, Fig. 50.*

Seed oval, rounded at one end, apiculate at the other, obscurely striate in the upper part, smooth.

The seed, evidently a nucleus, somewhat resembles a *Trigonocarpus.* But it has no distinct costæ and its base is round, polished without areolar scar. Its affinity is with *Rhabdocarpus plicatus*, Goepp., "Perm. Fl.," p. 170, Pl. XXVI, Fig. 1, of which the description applies to the American species. Comparing it with the figure, it differs by the striæ converging and gradually becoming stronger toward the point and gradually effaced toward the base or the rounded part of the seed which is perfectly smooth. The figure in Goepp. l. c., shows the fruit exactly oval and abruptly apiculate, while in our specimen it is rapidly narrowed into a short acute point but not abruptly so.

Habitat—Stanton mine near Wilkes-Barre, No. 979, Lacoe's collection.

TRIGONOCARPUS, *Brgt.*

Lesqx., "Coal Flora," p. 584.

TRIGONOCARPUS SCHULTZIANUS, *Goepp. & Berg., Pl. CX, Figs. 63–65.*

Seeds large, oblong, oval or ovate, parallelly costate, and very thinly striate on the pericarp between the ribs, truncate alveolate at the apex.

The specimens are numerous, some are covered with the pericarp others without it. It is not clear, however, if the thick sandstone layer which seems to cover the nucleus is not a part of the nucleus itself, the pericarp, generally transformed into coal upon fruits of this kind, being totally destroyed. The seeds in their whole or with what may be considered their pericarp, are 4 to 5 c. m. long, 1½ to 2½ c. m. broad, marked lengthwise by equal more or less acute or sharply prominent costæ, 11 or 12 in number, 6 of which are generally exposed upon the upper face. The minute striæ of the surface are rarely distinguishable. The species is related to *Rhabdocarpus multistriatus* Presl., as figured "Coal Flora," Pl. LXXXV, Figs. 22, 23.

Habitat—Communicated in a large number of specimens preserved in sandstone by Mr. W. H. Adams, from Stark county, Illinois, with the following species, also from Tracy, Tenn., and Stanton mine, Wilkes-Barre, No. 960, Lacoe's collection.

TRIGONOCARPUS ADAMSII, *Sp. nov., Pl. CX, Figs. 51–57.*

Nucleus ovate, acute or acuminate, sometimes enlarged on the sides by vertical compression, broadly alveolate at base, distinctly 6 to 8 costate, the costæ salient around the broad obtuse base or the alveole which is mamillate at the center, primary costæ separated by indistinct flat or obtuse striæ.

The seeds vary in size from 1 to 2 c. m. long and about 1½ c. m. broad. The nucleus only is preserved except in Fig. 53, where the border shows the width and the thickness of the ribs. In their original state, or with the outer testa, these seeds would be, by cross-section, represented as bearing six narrow wings 6 m. m. broad with the same number of intermediate narrow ones as in Pl. 23, Fig. 1 of Brgt. "Gr. Foss.," a species which the author describes as *Polypterospermum Renaultii.*

Habitat—Stark county, Illinois, with the last; also Stanton mine, Wilkes-Barre, and Port Griffith, Penna. No. 957 of Lacoe.

TRIGONOCORPUS PERPUSILLUS, *Sp. nov., Pl. CX, Figs. 58–61.*

Seeds very small, broadly ovate or globular, apiculate, rounded and alveolate, at the base, distinctly tricostate; surface smooth.

There are six specimens of this species, the largest 8 m. m. long, 5½ broad, the smallest 2½ m. m. long and about as broad. The alveole is seen Fig. 58.

Prof. Brongniart has figured and described in "Gr. Foss.," Pl. 22, Figs. 1–8, *Trigonocarpus pusillus*, a seed which merely differs from that of Illinois by its larger size and its ovate lanceolate-acuminate form, being 9 m. m. long and only 5 m. m. broad in its broadest part below the middle. The internal structure of the seed is represented by Brongniart in both the transverse and the longitudinal section.

Habitat—With the preceding, Stark county, Illinois. No. 971 of Lacoe's cabinet.

TRIGONOCARPUS GRANDIS, *Sp. nov.*, *Pl. CXI, Figs. 1–3.*

Fruit large, oblong-oval, rounded at both ends, alveolate, distinctly tricostate; costæ broad, flat; surface smooth, pericarp thick, inflated, truncate or sub-emarginate at apex.

These fruits, somewhat flattened, are 7½ c. m. long, 4½ c. m. broad, distinctly marked by three broad prominent costæ, and generally three shorter and narrower intermediate ones. The surface of one of the specimens, Fig. 3, is covered with a thick coating of brown coaly matter; representing the pericarp, truncate at the top, forming around the seed like an inflated ring, 5 m. m. in diameter.

The species bears a distinct relation to *T. Dawesii*, Ll. and Hutt., described below and abundantly represented in the same locality; but it is much larger, obscurely six costate, emarginate at one end, not pointed.

Habitat—Stark county, Illinois. No. 973 of Lacoe's collection.

TRIGONOCARPUS DAWESII, *Ll. and Hutt.*, *Pl. CXI, Figs. 4–6.*

Lesqx., "Coal Flora," p. 586, Pl. LXXXV, Figs. 2, 3, 25.

The specimens I refer to this species differ from it in some points and two of them at least, Figs. 4 and 6, may be small forms of the preceding. In these, the costæ are broad, but not prominent; the cicatrices or alveoles are round, not as large as in *T. Dawesii*. These differences are not very marked; but for Fig. 5, at least, the reference to the species appears certain.

Habitat—With the preceding. No. 664 of Lacoe.

TRIGONOCARPUS STARKIANUS, *Sp. nov., Pl. CXI, Figs. 7–13*

Seeds variable in size and aspect, generally broadly ovate, but also nearly round or oblong and narrow, narrowed upward to a truncate triangular point, tri-costate above, the costæ vanishing or becoming less distinct toward the base: rounded downward to a short generally broken pedicel, or sometimes flattened by compression.

These seeds vary from 1½ to 3 c. m. long, and from 1 to 2 c. m. broad. There are 18 specimens of this species, all showing more or less distinctly the scar or remnant of a pedicel at their base. In Fig. 11, the pedicel, apparently preserved in its whole length, is mamillate at its lower end.

Habitat—Stark county, Illinois, abundant. No. 972 in Mr. Lacoe's collection.

TRIGONOCARPUS KANSASEANUS, *Sp. Nov.*

Seeds rather small, ovate or oblong, marked by six more or less prominent costæ, the surface being smooth between them, pointed and trigonal at the apex, with a small rather pointed alveolar base; pericarp smooth, very obscurely costate, transformed into a pellicle of hard coal about ½ m. m. thick.

The seeds, 11 to 18 m. m. long, 7 to 9 m. m. broad in the middle, have great analogy to the small forms of the preceding species, and still more to those of *T. Adamsii*, described above, being only smaller, especially narrower in the middle, the basilar areoles smaller, less distinct, and the apex a trigonal short open acumen.

Habitat—Osage, Kansas. No. 791 of Lacoe's cabinet.

TRIGONOCARPUS MULTISTRIATUS, *Sp. Nov., Pl. CXI, Figs. 14, 15.*

Nucleus broadly ovate, abruptly acuminate or apiculate, tricostate, with thin intermediate striæ, parallel to the costæ; basilar alveole large, pointed in the middle.

The species resembles *Rhabdocarpus multistriatus*, Presl., "Coal Flora," Pl. LXXXV, Figs. 22, 23; but is a *Trigonocarpus* by its three strong costæ, the intermediate striæ being much thinner and not prominent. Both specimens represent a nucleus 2½ to 3 c. m. long, 2 to 2½ c. m. broad in the middle.

Habitat—These seeds were found among the specimens of coal plants of the National Museum, without indication of locality.

TRIGONOCARPUS AMPULLÆFORMIS, *Sp. nov. Pl. CIX, Figs. 18–20, 21?*

Seeds elliptical-oblong, narrowed upward to a tubulose neck, bordered by an enlarged or broad ring passing downward as a thick prominent costa.

These seeds appear costate, the borders, in Fig. 19, being like the enlarged or prominent ribs of *Trigonocarpus*, as it is more clearly seen Fig. 20, where on one side the costa passes down to the middle of the lower part of the nucleus while on the other it ascends along the neck like a border. Fig. 21 is covered with a coating of coaly matter, remains of the epicarp, and fragments of the ribs are seen in the upper part of the neck. Fig. 18 is hypothetically referred to the same species by its shape and an obscure ring formed by lines parallel to the borders converging under the neck and there joined in the form of a horse-shoe. These four specimens are fragmentary and seem to represent a species like *Cardiocarpus longicollis*, Figs. 24, 25 of the same plate.

Habitat—Sub-conglomerate coal of Arkansas, F. L. Harvey; Tracy and Rockwood, Tenn., No. 975 of Lacoe's collection.

CARPOLITHES. *St.*

Lesqx., "Coal Flora," p. 593.

The characters of this genus are not and cannot be defined. Most of the species are of uncertain relation. They represent mostly naked seeds some of them ovules or nuclei of seeds perhaps described in the three preceding genera.

CARPOLITHES CERASIFORMIS, *St., Pl. CXI, Fig. 18.*

Sternb., "Fl. d. Vorw.," II, p. 208, Pl. X, Fig. 9.

Seeds of medium size, round or oval, covered with an irregularly rugose pericarp.

These seeds apparently covered with a fleshy pericarp, not transformable into hard coal, are not rare in the lower coal measures. They are always flattened, 1 to 1½ c. m. in diameter, either quite round or broadly oval-obtuse with large wrinkles irregular in size and direction. The name given by Sternberg describes these fruits which are most like flattened dried cherries. They are most probably referable to some species of *Cordaites* and could therefore be described in the genus *Cordaicarpus.*

Habitat—At Cannelton and Arkansas. No. 999 of Lacoe's collection.

CARPOLITHES CONICUS, *Sp. nov., Pl. CXI, Fig. 17.*

A small seed, oblong, conical-pointed, rounded at base, partly flattened, convex on the surface; pericarp, a thin pellicle of hard coaly matter.

The small nutlet, a little more than 1 c. m. long, 6 to 7 m. m. broad in the middle, resembles the seeds of some *Cucurbitaceæ.* The pericarp is not more than ¼ m. m. thick.

Habitat—Shamokin, Pa. No. 959 of Lacoe.

CARPOLITHES BUTLERIANUS, *Sp. nov., Pl. CXI, Fig. 19.*

Seed flat, narrowly elliptical, obtuse at both ends, bearing fragments of a pericarp transformed into a thin pellicle of coal ½ m. m. thick and smooth like the surface of the nucleus.

This fine seed, 3½ c. m. long, 11 m. m. broad in the middle, is obscurely bordered by a narrow concave band and convex in the middle. Its relation is unknown.

Habitat—Butler mine E vein, Pittston. No. 990 of Lacoe's collection.

CARPOLITHES GRANULARIS, *St., Pl. CXI, Figs. 20, 21.*

Sternb., "Fl. d. Vorw.," 1, Pl. VII, Fig. 22.

Seeds exactly round, 5–6 m. m. in diameter, convex and smooth on the surface, either surrounded by a thick half terete narrow border, or globose only, and without border when compressed under a different plane; surface polished.

Habitat—Not rare. Boston mine, Pittston; also Osage City, Kansas. No. 981 of Mr. Lacoe's collection.

CARPOLITHES PERPUSILLUS, *Sp. nov., Pl. CXI, Figs. 22–24.*

Seeds very small, round or oval, obtuse, convex, generally less than 2 m. m. in diameter, the largest 5 m. m. long, 2 broad, all perfectly smooth but not polished.

Habitat—Sub-conglomerate, Campbell's Ledge, Pittston, Pa., not rare. No. 980 of Lacoe's collection.

CARPOLITHES MINIMUS, *St., Pl. CX, Figs. 66–68.*

Sternb. "Fl. d. Vorw.," Pl. VII, Fig. 3.

Seed small, 3½ to 4 m. m. in diameter, a little longer than broad, acute or acuminate at one end; surface either smooth or marked in the middle and longitudinally by a narrow costa which, on some of the specimens, appears to border the seeds as in species of Cardiocarpus.

These small seeds sometimes appear also obscurely tricostate much like *Trigonocarpus perpusillus*, Figs. 59 and 60 of the same plate. As the locality is different and the costa is either totally absent or scarcely distinct the species has its place with the *Carpolithes*.

Habitat—There are a large number of specimens of this species from Stanton mine near Wilkes-Barre and from Cannelton, all in the cabinet of Mr. Lacoe under No. 954.

CARPOLITHES LATIOR, *Sp. nov.*, *Pl. CX, Figs. 69, 70.*

Seeds much like those of the preceding species, but longer, oblong apiculate or mamillate at one end, rounded at the other, generally marked by one or two obscure striæ. As in the last species, these seeds are hard, not flattened but globular or cylindrical-ovate, showing their convex polished surface above the stone, but rarely separated from the imbedding matter. The pericarp is a thin coating of coal somewhat thicker than in *C. minimus.*

Habitat—Sub-conglomerate, Campbell's Ledge. No. 286 of Mr. Lacoe's collection.

CARPOLITHES TRANSSECTUS, *Sp. nov.*, *Pl. CXI, Figs. 27–27b.*

Seeds small, cut on both sides nearly to the middle by a rectangular aperture at right angles to the borders, rounded to the slightly truncate base, triangular, obtusely pointed, somewhat larger upward, narrowly margined all around, attached to flexuous thin flat smooth branches.

As seen Fig. 27*a* and enlarged 27*b*, a seed separated from the stem, as they are mostly found upon the shale, these bodies, 6 m. m. long and as broad, have a pentagonal-like outlined form marked both sides by rectangular incisions penetrating to near the middle. The surface is concave, polished; the border narrow, convex, not often distinct; the point of attachment, Fig. 27, is not positively seen, the seeds appearing either fixed to narrow stems by the middle or by what appears to be the apex. They are strewn upon the stone, generally loose and very irregularly disposed, mixed with fragments of stems which are found upon all the specimens. These organs may be the fructifications of a fern. I do not know that anything of the kind has been observed before.

Habitat—Sub-conglomerate coal of Arkansas, F. L. Harvey. No. 279 of Lacoe's collection.

ADDITIONS AND CORRECTIONS.

A number of the species recorded in the following Table of Distribution have not been described as yet in the Coal Flora for the following reasons.—I do not like to give for publication descriptions of fossil plants when the species cannot be illustrated by figures. Since the plates of this volume have been prepared, I have had for examination a great number of specimens, some of them representing new or interesting species which were left out for future publication, in case the number of plates could be increased if the Survey was continued. Some species, also, insufficiently known had been omitted until they could be positively determined by comparison, if better materials were obtained. As I have now to definitely close the work, I give here, in order to have in it as complete a representation as possible of the carboniferous vegetation of the United States, short descriptions or references to authors, of all the species of fossil plants, which, recorded in the following table, have not been mentioned in the Coal Flora :

THALASSOPHYTES OR MARINE ALGÆ.

DICTYOPHYTUM, HALL.

16th Ann. Rep. of the Cabt. of Natural History of New York, p. 87.

Fronds flabellate or funnel-shaped, latticed by radiate and concentrical lines crossing at right angles ; stems obconical or subcylindrical, hollow, striate outside, equal, or articulate and nodose at the articulations.

DICTYOPHYTUM RAMOSUM, *Sp. nov.*

Stem nodose or tubercular at the articulations, the tubercles projecting outside at right angles to the stem, as short branches, inflated and button-like at the apex.

The species may be a form of *D. tuberosum*, Hall, l. c., p. 90, Pl. III, Fig. 1, which has the articulations inflated into short obtuse tubercles, some of them broken, but none prolonged as branches. The specimens I have had for examination represent the tubercles arranged in horizontal rows, as in *D. tuberosum*, but instead of being round and obtuse they are prolonged horizontally into short branches inflated into half round knots at the apex. The ribbon-like divisions surrounding the stems and branches are parallel and quite distinct.

Habitat. Upper Chemung, head of Hill creek, Charleston township, Tioga county, Penna. A. Sherwood.

CALAMARIEÆ.

MACROSTACHYA, *Schp.*

"*Coal Flora*," p. 60.

I have described under this name spikes of various sizes composed of a comparatively large axis, to which are attached lineal scales, connate at base or in the lower part and diverging first at right angles from the axis, then abruptly bent upward from the middle and imbricated, separated and acuminate at the apex. Pl. III, Figs. 17–20. As remarked in the description, these spikes have a degree of analogy to those of *Volkmannia* and apparently are similarly conformed, as seen in *Volkmannia fertilis*, Pl. XC, Fig. 4, differing merely, in appearance at least, by the narrow central axis.

Of the spikes of *Macrostachia*, we have, as yet, three different forms or species to be named.

MACROSTACHYA COMMUNIS, *Sp. nov., Pl. III, Figs. 17–18.*

Spikes large, oblong, obtuse, truncate or pedicellate at base, scales large, lanceolate-acuminate at apex, generally broken or truncate above.

These spikes, considered by Schimper as those of *Macrostachya infundibuliformis, Schp.*, a fragment of which is figured Pl. III, Fig. 14, are locally very abundant, generally mixed with stems and branches of *Asterophyllites equisetiformis.* They are generally erect, but sometimes also curved or scythe shaped.

Habitat. Cannelton; West-Wood, near Pottsville, abundant; Archibald and Olyphant, Penna., etc.

MACROSTACHYA APERTA, *Lesqx., Pl. III, Fig. 20.*

Asterophyllites aperta, Lesqx., "Geol. of Penna., 1858," p. 852, Pl. I, Fig. 5.

Spike long, linear, gradually narrower toward the base; bracts or scales open, horizontal, cut from the middle, exposing the broad striated axis and the space between the articulations.

These spikes evidently belong to the same genus as the last ones described, though they are far different in appearance. They are narrower, linear, comparatively longer and the scales are generally open and truncate.

Habitat. Rarely found. New Philadelphia, coal M, and Cannelton, Penna.

MACTOSTACHYA MINOR, *Sp. Nov., Pl. III, Figs. 19, 19 a.*

Spikes short and narrow, linear or narrowly oblong, abruptly pointed; scales closely imbricate and appressed, lanceolate-acuminate from the middle upward, narrowed at base into a short pedicel.

At first I considered this form as a mere variety of *M. communis.* But the species has been found at the same locality in numerous specimens, all with the same character and of the same size. The spikes are generally narrower, more linear than the one figured.

Habitat. Campbell's Ledge, near Pittston, interconglomerate. No. 212 of Mr. Lacoe's collection.

FILICACEÆ.

CYCLOPTERIS, *Auct.*

CYCLOPTERIS VALIDA, *Daws.*

"Dev. Plants," Quarter. Journ. Geol. Soc. XVIII," p. 319, Pl. XVII, Fig. 52.

Tripinnate; primary divisions of the rachis stout and wrinkled. Pinnæ regularly alternate. Lower pinnules nearly as broad as long, deeply and obtusely lobed, narrowed and decurrent at the base, regularly diminished in size and breadth toward the point, and the last pinnules narrowly obovate and confluent with the terminal pinnule. Nerves delicate, several times dichotomous. (*Dawson.*)

Similar in all its characters to *Pseudopecopteris macilenta*, Ll. & Hutt., from which it differs, says the author, by essential characters (not mentioned.)

Habitat. Upper Devonian, Pennsylvania. In the table of the Dev. plant l. c. the species is marked only as found in New Brunswick.

CYCLOPTERIS INCERTA, *Daws.*

"L. c.," p. 320, Pl. XVI, Fig. 44.

Stipes 1 c. m. in diameter, giving off branches at acute angles, on which are borne below, branching and re-curved remains of fertile pinnules, and above, traces of terminal obovate pinnules. (*Dawson.*)

An undeterminable fragment.

Habitat. Hamilton group of New York.

ODONTOPTERIS, *Brgt.*

"Coal Flora," p. 124.

ODONTOPTERIS BRITANNICA, *Gutb.*

Abdr., p. 68, Pl. 9, Fig. 8-11.

Frond polypinnate; rachis narrow; ultimate pinnæ open, nearly at right angles; pinnules close, short, oblong, obtuse, the lower connate at the base, the upper gradually shorter and connate to the middle; medial nerve distinct; lateral veins partly derived from the main rachis, partly from the midrib, all distinct.

The specimen from which the description is made is like a counterpart of that represented Figs. 8 and 9, Gutb., l. c., and also like that in Weiss' "Flora Foss.," Pl. I, Fig. 2. The figures, as also the specimens I have seen, are entirely different from any of those of *O. Schlotheimii*, to which Schimper refers *O. Britannica* as a synonym.

Habitat. Cannelton, Pennsylvania; St. Clairsville, Ohio. No. 753 Lacoe's collection.

ODONTOPTERIS REICHIANA, *Gutb.*

Abdr., p. 65, Pl. I, Fig. 1-5, 7; X, Fig. 13.

Fronds polypinnate and sometimes dichotomous at apex; ultimate pinnæ open, linear-lanceolate, rigid or flexuous; pinnules narrowly ovate-lanceolate, from a truncate base, obtusely pointed, the upper narrower the terminal longer, lanceolate, obtusely acuminate from an enlarged base.

Of the specimens which I refer to this species, one has the pinnules oblong, somewhat enlarged at the base, inclined upward, blunt-pointed, 15-17 m. m. long, 1 c. m. broad at base, the lateral veins emerging from the rachis, the medial nerve distinct and dichotomous above, the veins being distant, thin, slightly curved toward the borders, forking once, rarely twice. The other specimen has the leaflets smaller, broader at base, ovate-acute, the lower pinnules enlarged and lobed, as in Fig. 2 of Gutb., l. c.

Habitat. Thayer, and Osage City, Kansas. No 754 of Lacoe's collection.

ODONTOPTERIS DILATATA, *Sp. nov.*

Terminal pinnule broadly ovate or obovate, obtuse, the lateral oval or oblong, obtuse, decurring and connate at base by a narrow band bordering the rachis; veins close, dichotomous, the lateral slightly curved outside; texture thick; surface polished.

A mere fragment compared to *O. Obtusa*, Brgt., but without distinct affinity to any species known to me. The fragment is too small for a satisfactory diagnosis.

Habitat. Cannelton, Pennsylvania.

LESLEYA, *Lesqx.*

"Coal Flora," p. 143.

LESLEYA MICROPHYLLA, *Sp. nov.*

Leaves oval-oblong, equally narrowed upward and downward from the middle, obtuse or truncate at the apex, joining the petiole in decurring or by a small auricle, petiolate; lateral nerves all emerging obliquely from a thick convex medial nerve, curving in passing to the borders, di chotomous, forking two or three times.

I have seen two leaves of this species, one 4 c. m. long, 16 m. m. broad in the middle, shortly auriculate at base, with a broken petiole 7 m. m. long. In both specimens the lateral veins joining the medial nerve at a very acute angle of divergence reach the borders obliquely at an angle of about 50°, and count there 40 per c. m. The other leaf merely differs from the first in the basilar borders without auricles. It is better preserved in its whole length, being truncate at the apex ; in the other specimen the top of the leaf is destroyed.

Habitat. Lawrence and Osage, Kansas. No. 314 of Mr Lacoe's collection.

DICTYOPTERIS, *Gutb.*

"*Coal Flora,*" *p. 143.*

DICTYOPTERIS SCHEUCHZIRI, *Hoffm.*

"*In Roem., Pflanz. d. Kohlengeb. am Harze, Palæont.*" *IX, p. 186, Pl. XXXII, Fig. 1.*

Pinnules large, nearly sessile, cordate or truncate at base, ovate-lanceolate, the terminal larger, oblique; medial nerve indistinct; lateral veins anastomosing in their flex ures, forming large, rhomboidal meshes.

I refer to this species two specimens of different aspect. One is a single leaflet, long, nearly linear, rounded at apex, with lateral nerves prominent or filiform, not flattened as in *D. rubella*, and thus very distinct, polished. The other is a fragment of a pinna with three pinnules of different sizes, the smallest only 14 m. m. long, 5 m. m. broad, the largest 9 to 10 c. m. long, 6 c. m. broad, all ovate, cordate at base.

The leaves of the second specimen are all fragmentary Though the nervation is the same, they may represent a different species, the first leaflet described being comparatively long and narrow, 1 c. m. broad at base and in the middle, with borders flexuous.

Habitat. This last specimen, a single leaf, is from Port Griffith, the other from Moosic, Penna. Anthracite Coal Co.'s mine. Nos. 823 and 823 c of Lacoe.

DICTYOPTERIS CORDATA, *Roem.*, *l. c.*

Pinnules cordate at base, obliquely acuminate, repand-sinuous on the borders; lateral nerves close, dichotomous, anastomosing in their curves and forming large rhomboidal meshes.

The species resembles *Neuropteris hirsuta*, Lesqx.; and may be a derived form of it, as it has sometimes the veins somewhat undulate, but I have not seen any leaves of this last species with anastomosing flexures.

Habitat. Nodules of Mazon creek, shale over coal at Morris, Ill. No. 631 Lacoe's collection.

DICTYOPTERIS NEUROPTEROIDES, *Roehl.*

"*Foss. fl. des Steinkohlen form. Westphal.*," *p. 49, Pl. XIV., Fig. 6.*

Lateral leaflets close, slightly scythe shaped, imbricate on the borders or contiguous, oblong-lanceolate, obtuse, nearly truncate at base, the upper basilar border only rounded, not auricled at base; upper terminal pinnules comparatively long, linear-lanceolate, obtuse.

By the size and form of the leaflets, this species resembles *D. Brongniarti*, but the lateral nerves are more distinct and the meshes longer and narrower. The pinnules are nearly 2 c. m. long and 8 m. m. broad at base; the medial nerve is broad, flat, gradually narrower upward, vanishing under the apex; the veins are thick, forking generally three times, anastomosing near their base and in the middle, but arched and free toward the borders. The size of the pinnules is variable upon the same pinna.

Habitat. Morris, Ill. No. 633 of Mr. Lacoe's cabinet.

MEGALOPTERIS, *Daws.*

"*Coal Flora*," *p. 147.*

MEGALOPTERIS DENTATA, *Sp. nov.*

Pinnules lanceolate, acute, larger above the middle, gradually narrowed toward the decurring base, dentate on the borders, perhaps by compression or local abrasion of the thick substance of the leaves when softened by maceration.

Specie uncertain. It is represented only by the top of a pinna with 4 leaflets, 4 to 5½ c. m. long, 1 c. m. in diameter at the broadest part, 5–6 m. m. above the decurring base. The veins are very oblique, distinct and distant, being only 20 per c. m. along the borders. The medial nerve is indistinct.

Habitat. Rushville, Ohio. No. 783 of Mr. Lacoe's collection.

MEGALOPTERIS SERRATA, *Sp. nov.*

Pinnules narrow, at an acute angle of divergence, narrowly lanceolate, long sharply acuminate, sometimes bifid above the decurring base; veins strong and distant, very oblique, their points often passing out of the borders which thus appear serrate or denticulate, but with much smaller teeth than in the preceding species.

The leaflets, 2–4 c. m. long, are only 5 m. m. broad. The whole plant resembles a *Rhacophyllum*.

Habitat. Rushville, Ohio. No. 789 of Lacoe's cabinet.

Megalopteris Sewellensis, Fontaine, is a species provisionally named but not described by the author in "Jour. Sci. and Arts," 3d Series, XI, p. 383.

ALETHOPTERIS, *St.*

"*Coal Flora*," *p. 175.*

ALETHOPTERIS EVANSII, *Sp. nov.*

Ultimate pinnæ narrow, linear, pinnately lobed; lobes obtuse or triangular, obtusely pointed, short, connate to the middle, each marked with a distinct medial nerve; lateral veins nearly erect, thick, distinct, those emerging from the rachis erect or even sub-incurved, those above, derived from the medial nerve in a very acute angle of divergence, close, forking once.

The species resembles *A. Pennsylvanica*, differing by the form of the pointed lobes and the nervation. The specimens are, however, too small, though very numerous, for a sufficient diagnosis.

Habitat. Dade, Georgia; also Tennessee. No. 1013 of Lacoe.

ALETHOPTERIS ROBUSTA, *Sp. Nov.*

Pinnæ large, pinnately divided; ultimate pinnæ or pinnules at right angles, distant, rigid, linear, long and narrow, medial nerve very thick.

The ultimate pinnæ appear like long narrow linear pinnules, the broad medial nerve or rachis being bordered with a narrow, undulate lamina; sometimes the lamina, slightly broader, is marked by deeper undulations or short quadrangular lobes, 4 m. m. long, as broad at base, with a thick medial nerve continuous to the apex, the lateral veins being simple and oblique. The substance is very thick, the texture coarse. This renders the diagnosis as yet insufficient.

Habitat. Cannelton. No. 743 of Lacoe's cabinet

PECOPTERIS, *Brgt.*

"Coal Flora," p. 223.

PECOPTERIS SCHIMPERI, *Font. and White.*

Perm. or Upper Carb. Flora of West Virginia, p. 75, Pl. XXIV, Fig. 1-5.

Frond tripinnate; primary rachis stout and rather rough; secondary pinnæ alternate, linear-lanceolate, taper pointed, going off at nearly right angle; pinnules short, ovate, or triangular in outline, alternate, the lower united near the base, the upper gradually more connate, becoming confluent near the apex; texture thick and leathery; mid-nerve strong, flexuous, reaching the apex; lateral nerves stout, dichotomously forking, the branches all deflexed, joining the borders nearly at right angle. (Font. and White.

Habitat. Great vein of the Hocking coal fields, Ohio, horizon of the Pittsburgh coal, numerous, in fine specimens. Lacoe's, No. 886.

SPHENOPTERIS, *Brgt.*

"Coal Flora," p. 269.

SPHENOPTERIS CRENATA, *Ll. & Hut.*

"Foss. Flor.," Pl. C, CI.

Fronds ovate or oblong-lanceolate, tripinnate; pinnæ subopposite and alternate, open, subarcuate, long-lanceo-

late; pinnules subopposite, linear-lanceolate, subpinnatifid or crenulate; nerves obsolete.

Habitat. Coal measures of Arkansas; F. L. Harvey.

SPHENOPTERIS TENELLA, *Brgt.*

"*Veget. Foss.*," *p. 186, Pl. XLIX, Fig. 1.*

Pinnæ oblong-lanceolate, bipinnatifid; secondary pinnæ oblique, rigid, oblong or broadly lanceolate, close and parallel, with 5 or 6 rows of alternate pinnules, open, sessile pinnately divided into 2 or 3 pairs of alternate simple linear obtuse entire short lobes obliquely diverging.

All the lobes are entire, except the lowest on the upper side of the pinnules which is generally a little enlarged at the apex and there truncate, emarginate or biffd. The lobes are open, not decurrent; the medial nerve is pinnately divided, one of its branches entering into each lobe and ascending to the apex. In Brongniart's figure, the lateral pinnæ are shorter, the pinnules close, the lobes entire, none of them bifid at the apex. But the author says that the species is known to him only by a drawing and he considers it as different from *S. tridactylites.* From the last species, the American form described above is very different by the narrow filiform entire lobes, etc. It is more like *S. elegans,* Brgt., from which it differs also by the narrower longer simple lobe, narrow rachis, etc.

Habitat. Thayer, Kansas. No. 254 of Lacoe's collection Horizon not ascertained yet.

SPHENOPTERIS DISSECTA, *Brgt.*

"*Veget. Foss*," *p. 183, Pl. 49, Figs. 2, 3.*

Bipinnatifid, rachis winged; pinnules deeply pinnatifid; divisions linear, very narrow, distant, distinct to the base, simply nerved, the lower bi or trifurcate; lobes obtuse, diverging.

The pinnæ are oblique, parallel, alternately pinnately lobed, lobes distant, the lower trifid, the middle bifid, the upper entire, all long and linear.

Habitat. Old Forge township, Carbon Hill mine, coal D, Lacoe's collection. No. 371.

Sphenopteris fascicularis, *Roem.*

"Beitr. in Paleont.," IX, p. 179. Pl. XXVIII, Fig. 9.

Pinnæ with close alternate oblique ovate pinnules, cut in many close or connate laciniæ, becoming free and distinct on the borders and more or less curved back.

The species is difficult to define and as yet I have never seen it distinctly, no more than it is seen upon the figure given by the author. The pinnules ovate or oblong-acuminate, sessile upon a narrow rachis, appear as composed of nerves of irregular size whose points curving near the borders, attenuated and free, pass out of them like decomposed bundles of filaments. The species may be referable to the genus *Rhacophyllum*. The borders of the pinnules are not defined by any line.

Habitat. Olyphant, Penna. No. 367 of Mr. Lacoe's collection. Also found at Cannelton.

Sphenopteris Hitchcockiana, *Daws.*

"Flora of the Devon. Per. of N. E. Am., Quart. Jour. Geol. Soc.," Vol XVIII, p. 321, Pl. XVI, Fig. 51.

Stipes stout, straight, rugose, giving off slender secondary petioles which ramify dichotomously and terminate in minute ovate leaflets. (Dawson.)

This plant, as figured, appears to represent fruiting branches of some species of *Archæopteris*.

Habitat. Devonian of Perry, near Eastport, Maine.

Archæopteris, *Daws.*

"Coal Flora," p. 299.

Archæopteris (Cyclopteris) Brownii, *Daws.*

L. C., p. 320. Plate XII, Fig. 9.

Pinnules large, connate, with distinct once-forked nerves and waved margins. (Daws.)

A small fragment of uncertain generic attribution.

Habitat. Same as the last.

Rhacophyllum, *Schp.*

"Coal Flora," p. 309.

RHACOPHYLLUM FILIFORME.

Fucoides filiformis, Guth., "Abdr.," p. 12, Pl. 1, Figs. 9, 9a.

Primary rachis somewhat thick, alternately dichotomous; primary branches at an acute angle of divergence, dichotomously divided from the base into thin, thread-like branches diverging also in acute angle, the upper branches becoming gradually more slender.

The main stem or rachis is 4 m. m. in diameter; the secondary 2 m. m.; the ultimate divisions ½ to 1 m. m. The divisions vary in length, and are sometimes 5–6 c. m. long.

Habitat. Rhode Island coal beds; Garnet, Kansas. No. 813 of Mr. Lacoe's.

STEMMATOPTERIS, *Corda.*

"Coal Flora," p. 337.

STEMMATOPTERIS MICROSTIGMA, *Sp. Nov.*

Scars small; borders broad, parallel, not emarginate at top; surface of the inside scars punctulate.

Similar to *S. punctata;* the scars are only half as long, though of the same width.

Habitat. Cannelton, Pennsylvania. No. 612 of Lacoe's collection.

STEMMATOPTERIS ANCEPS, *Sp. Nov.?*

Scars broadly oval, with a broad parallel border; vascular scars horse-shoe-shaped; surface dull, smooth.

Species made from two indifferent specimens and uncertain.

Habitat. Cannelton. Pennsylvania; No. 611 of Lacoe's collection.

RACHIOPTERIS, *Daws.*

"Coal Flora," p. 331.

RACHIOPTERIS SQUAMOSA, *Sp. Nov.*

Rachis or rhizoma cylindrical, flattened by maceration, irregularly pustulate or marked by close elevated conical points irregular in position, representing the base of long scaliform hairs; hairs or scales linear-lanceolate, curved up or hooked, bordering the stems when they have not been destroyed by maceration.

These stipes are probably those of *Pecopteris villosa* or *P. squamosa*.

Habitat. Not rare at Cannelton. No. 456 of Lacoe.

LYCOPODIACEÆ.

LEPIDODENDRON.

"*Coal Flora,*" *p. 363.*

LEPIDODENDRON RIGIDUM, *Sp. Nov.*

Branches small; leaf-scars distinct, broadly rhomboidal; leaves thick, coriaceous, very rigid, polished, gradually narrowed to a very sharp point; medial nerve broad, convex.

I have not seen other leaves of the same character. They are 8–10 c. m. long, 4 m. m. broad at the base, thence gradually tapering upward to a sharply acute point. These leaves somewhat resemble those of *L. rigens*, Lesqx., being, however, flat, only slightly concave toward the base; the lower leaves are more distant, at right angles to the stem; the upper oblique, turned upward, not crowded.

Habitat. Dade mine, Georgia; also sent from Tennessee. Fig. 918 of Lacoe.

KNORRIA, *St.*

"*Coal Flora,*" *p. 407.*

KNORRIA SELLONI, *St.*

"*Flora d. Vorw.,*" *1, p. 37, Pl. LVII.*

Cortical scars long-rhomboidal; bolsters distant, conic-linear, truncate at the apex.

Differs from *Knorria imbricata* by the bolsters distant and narrower.

KNORRIA COMPACTA, *Sp. Nov.*

Branches or trunks small, covered with appressed leaves apparently tubulose or inflated a distance above the base, then flat, long, linear, with a broad flat mid-rib; leaf-scars close, in spiral order, angularly round or round-

hexagonal, marked in the middle by an elevated small rarely distinct mamilla.

The species is represented by three specimens, two of which are apparently fragments of trunks, flattened and without leaves; the other is a branch 5–6 c. m. in diameter, 15 c. m. long, bearing leaves and entirely covered upon the exposed surface by scars of the leaves which still remain attached to the borders. The scars measure in diameter 5 m. m.; they are convex or more elevated in the middle, where one may sometimes discern a small mamilla, but more generally they are merely convex and smooth. The outlines of the scars, which are quite close, (2–3 m. m. distant,) show the tubulose base of the leaves which, as seen along the borders of the branch, are inflated or tubulose at base for 2 c. m. of their length, and then flat, 5 m. m. broad and long, as far as can be seen from the specimen which they cover its whole length to the broken borders, compressed, canaliculate in the middle by a broad flat nerve more than 1 m. m. broad. Upon the trunks, the scars are surrounded at base by a narrow more elevated border.

The reference of this peculiar plant to *Knorria* is not positive.

Habitat. Thayer, Kansas. No. 816 of Mr. Lacoe's collection.

LEPIDOSTROBUS.

"*Coal Flora*," p. 451.

LEPIDOSTROBUS BUTLERI, *Sp. Nov.*

Strobile oblong or narrowly oval, comparatively short; bracts erect, imbricate and appressed, long, distinct, linear-lanceolate.

The strobile is 9 c. m. long, 3½ c. m. broad in the middle, and flattened; the bracts 2½ c. m. long, crowded and appressed, 2 m. m. broad near the base, are gradually acuminate and broadly nerved. The species closely resembles *L. variabilis*, Ll. & Hut., differing essentially in the bracts longer, more distinct, the strobile oblong or oval. One of the specimens, No. 681*a*, is an intermediate form with bracts more distinct or coriaceous, and broader than in *L. variabilis*, but shorter and narrower than in *L. Butleri*.

Habitat. Butler mine, near Pittston, Pennsylvania. No. 681 of Lacoe. No. 681*a* is from Brown colliery, near Pittston, and in same collection.

LEPIDOSTROBUS LATUS, *Sp. Nov.*

Bracts long, lanceolate, erect, sporangiophores oblanceolate, long and gradually narrowed to the base.

It is a fragment of strobile, 9 c. m. long, 5 c. m. broad, with bracts 3½ c. m. long, 5 m. m. broad at base, attached to a long slender oblanceolate sporangiophore 1½ c. m. long. The blades, by their shape and length, are like those of *L. lanceolatus*, Brgt., somewhat longer and comparatively narrower; but in this last the sporanges and sporangiophores are short, not half as long as those of the species. By this last character it is much like *L. prælongus*, Lesqx., but the sporanges are at right angle, not inclined upward.

Habitat. Olyphant; No. 728 in Mr. Lacoe's collection.*

LEPIDOPHYLLUM, *Brgt.*

"Coal Flora," p. 447.

LEPIDOPHYLLUM STANTONI, *Sp. Nov.*

Blade lanceolate from a broad base horizontally prolonged into two acute auricles; sporange joined its whole width to the base of the blade or the auricles, abruptly narrowed a little below to the sporangiophore; blade with a broad flat obsolete nerve.

The species essentially differs from *S. hastatum* by the shorter blade broadly lanceolate acuminate abruptly enlarged at base into acute auricles, and by the broad sporange also abruptly narrowed to the sporangiophore and somewhat longer.

Habitat. Stanton mine, near Wilkes Barre, Penna. No. 657 of Lacoe's cabinet.

**Lepidostrobus Richardsoni*, Daws., and *L. globosus*, Daws., are quoted by the author in "Devon. Plants. Quater. Journ. Geol. Soc.," XVIII, p. 314; both from the Devonian of Maine.

SIGILLARIEÆ.

SIGILLARIA, *Brgt.*

"*Coal Flora*," *p. 467.*

SIGILLARIA BROCHANTI, *Brgt.*

"*Veget. Foss.*," *p. 442, Pl. CLIX, Fig. 2.*

Stem narrowly costate, undulate or alternately contracted, smooth; scars obliquely rhomboidal-lanceolate, acute at the angles, broadly obtuse or arcuate in the upper and lower part; vascular scar central, punctiform.

Habitat. Butler mine, coal E, Pittston, Penna. No. 618, Lacoe's cabinet.

SIGILLARIA SAULLII, *Brgt.*

"*Veget. Foss.*," *p. 456, Pl. CLI.*

Ribs narrow, undulate, separated by deep, narrow furrows, areoles round-oval, as broad as long, somewhat narrowed, truncate or convex at the top, rounded at the base, slightly angular at the sides, somewhat distant and separated by transverse wrinkles; vascular scars three, placed in the upper part of the areoles.

Habitat. Plymouth, Penna. No. 471 of Lacoe's collection.

SIGILLARIA VENOSA, *Brgt.*

"*Veget. Foss.*," *p. 424, Pl. CLVII, Fig. 6.*

Stem smooth (not ribbed), flat; cortex thin, longitudinally striate or veined; areoles placed in quincunxial order, large and distant, broadly ovate, truncate or emarginate at the upper border, rounded at the lower, enlarged below the middle; vascular scars three, placed in the middle of the areoles.

The areoles, about 1 c. m. long, 8 m. m. broad, are 3½ c. m. distant obliquely, and 4 c. m. vertically. The characters of the American form do not differ from those indicated by Brongniart; the wrinkles or longitudinal striæ are only more distinctly marked in the specimen I have examined.

Habitat. Plymouth, vein F. Lacoe's, No. 900.

STIGMARIA, *Brgt.*

"*Coal Flora,*" *p. 509.*

STIGMARIA ELLIPTICA, *Lesqx*

"*Geol. Rep. of Ill.,*" *IV, p. 451, Pl. XXIX, Fig. 2.*

Stem thick, flattened by compression; scars or areoles placed in regular spiral order, distant, elliptical, blunt or acutely pointed, with a central small nearly round mamilla, marked in the middle by a vascular point.

The specimen is covered with a thin coating of coaly matter which has filled the concave areoles, obliterating the mamilla which is only distinguishable at the bottom of some scars. The specimen may represent some of the varieties ascribed by authors to *Stigmaria ficoides*, Brgt.

STIGMARIA EXIGUA, *Daws.*

"*Geol. Surv. of Can., Fos. Pl.,*" *p. 23, Pl. III, Fig. 30.*

Scars small, in depressed spaces, six in 2½ c. m., vertically; stem cylindrical, 2½ c. m. in diameter. (*Daws.*)

The author remarks that it is questionable whether this is not a branch of a species of *Cyclostigma* rather than a *Stigmaria.*

Habitat. Upper Devonian, Elmira, New York.

NOEGGERATHIEÆ.

WHITTLESEYA, *Newby.*

"*Coal Flora,*" *p. 523.*

WHITTLESEYA MICROPHYLLA, *Sp. Nov.*

Leaves or leaflets small, broadly obovate, or flabelliform, rounded-subtruncate at top, narrowed from the middle downward by a slight curve to a short petiole; veins straight, apparently dichotomous, irregular in thickness; epidermis thickish, smooth.

The epidermis or upper surface is a pellicle of coal only partly preserved, and under which the nervation is indistinctly seen. The leaflets without the petiole are a little more than 1 c. m. long and 10 m. m. broad in the upper part. By their form and nervation these small leaves have an affinity to those of *W. integrifolia*, Lesqx.

Habitat. Arkansas coal measures. F. L. Harvey.

CORDAITEÆ.

"Coal Flora," p. 525.

CORDAIANTHUS, *Grd' Eury.*

CORDAIANTHUS SCABER, *Sp. Nov.?*

Differs merely from *C. rugosus* by the rachis which is not transversely rugose, but lineate lengthwise and rough, punctulate as by short hairs.

CORDAIANTHUS EBRACTEATUS, *Sp. Nov.*

Stem slender, narrow; nutlets opposite, distant, small and globose, apparently composed of obtuse appressed scales, without bracts underneath.

The ovules and scales are not distinct; the specimen is not good enough for positive determination.

RHABDOCARPUS, *Goepp. & Berg.*

"Coal Flora," p. 574.

RHABDOCARPUS BEINERTIANUS? *Goepp. & Berg.*

"De fruct. et. Semin.," p. 20, Pl. 1, Figs. 8 (on the left), 9.

Seeds oblong, narrowed at base, surrounded by an undulate margin, marked by parallel close thin lines or nerves, and covered by a thick testa.

Habitat. Arkansas; No. 797 Lacoe's cabinet.

RHABDOCARPUS BOCKSCHIANUS, *Goep. & Berg.*

L. c., p. 21, Pl. I, Figs. 13 and 14.

Seeds elliptical, narrowed upward into a long, styliform

acumen ; very thinly striate, pointed or truncate at base, often covered with a thick epidermis.

The description of these fruits is made in the supposition that the figures of the author are overturned, the basilar prolongation of Fig. 14 being like the long neck or micropyle of the fruits which I have described and figured as *Cardiocarpus longicollis*, p. 808.

Habitat. Arkansas coal measures, F. L. Harvey. Rockwood, Tennessee. No. 1029 Lacoe's collection.

Geographical and Stratigraphical Distribution of the Plants Described in the Coal Flora.

From the beginning of my researches on the vegetation of the carboniferous of the United States, I have endeavored to find if, as I supposed it at first, and as it was surmised by European authors, it would be possible to recognize, by the presence of certain species of plants, the divers horizons of the coal strata in different localities. As long as the observations were confined to a field of limited area, the plants found in connection with coal beds of the same horizon did, indeed, present so great an affinity by specific identity that the possibility of ascertaining the horizon of a coal from the characters of its plants appeared most probable. In the shale over the coal of St. John, Illinois, for example, I found only 7 species, 4 of them identical with those of the coal bed of DuQuoin, worked at two miles distance. The remains of plants also of the vein of coal named Salem or Gate, near Pottsville, Mount Carbon, New Philadelphia, and Tremont, Pennsylvania, present the same analogy.

But against these and other few cases of marked identity of the plants of certain beds of coal, it is right to say that generally the data furnished by remains of fossil plants have proved to be either insufficient or deceptive. Insufficient, because the specimens of fossil plants are rarely found representing such a number of species that a satisfactory comparison may be made between the plants of different localities ;—deceptive, because the vegetation of the coal period like that of the present epoch has been subjected to great variations, according to geographical distribution or local influences. Therefore, the plants found in connection with two beds of the same horizon may be far different in types, especially when observed at distant localities. And per contra, it has been seen also that sometimes two beds of coal at different levels have been formed of plants of closely allied even identical species.

What it is until now possible to clearly see, in the vertical distribution of the plants of the coal measures, is that the types of the pre-carboniferous and carboniferous flora have been subject to four distinct modifications of their essential characters, and that therefore the series is composed of four distinct groups generally marked upon the whole extent of the formation.

The first comprising the lower or pre-carboniferous, is persistent in the Devonian, the Chemung, and the Catskill.

The second, partly marked in the Pocono, becomes modified and persistent in the sub and intra conglomerate measures.

The third pertains to the lower productive coal measures from above the conglomerate to the base of the barren beds of Pennsylvania.

The fourth is that of the upper productive coal measures, including the Pittsburgh coal and a few hundred feet above. Of course, some species of each group have a greater degree of persistence and pass from a lower to an upper stage.

In the 2nd volume of this work, p. 669, Chapter VI, I have said, on the stratigraphical distribution of the coal plants most of what is known until now on a general point of view. The remarks have been derived from a table of distribution which was composed from the materials I had then obtained. Now, though I have nothing to add to those remarks, a number of new species having been described since in the third volume, it would be advisable to have them inserted with the others, in a table of the same kind, for future comparison.

But the first table, if it does represent well enough the stratigraphical distribution of most of the species, does not take into account all the localities where the plants have been found, and this is a very important matter to be considered, the localities giving, by the number and characters of their plants, fixed points for comparison and researches.

As it is not possible to name all the localities upon a table like the one formerly made, I have prepared, to close the volume, a simple table of distribution by a number of columns about like those of the first, indicating the geological horizon either known or supposed where each species has been recognized.

The number of places where a species has been found is merely marked by numbers in each column.

Besides this I have made lists of all the species of coal plants with the names and horizons of the localities where each of them has been found.

The table, therefore, and the enumeration of the plants will afford the means of comparing both the geographical and stratigraphical distribution of the plants, while at the same time, they give, taken together, a clear exposition of the vegetation of the carboniferous period of the United States, as far as it is known to the present time.

The present Table of Distribution is composed of 17 columns.

Column 1st. Devonian.
" 2nd. Chemung.
" 3rd. Catskill.
" 4th. Pocono and sub-carboniferous.
" 5th. Sub-conglomerate measures.
" 6th. Intra-conglomerate measures.
" 7th to 11th. Anthracite productive measures, viz:
7th. Coal A, B, C.
8th. Coal D, E, F.
9th. Coal G to M.
10th. Coal M and above.
11th. Coal of unknown horizon.
" 12th to 16th. Bituminous productive measures, viz:
12th. Coal A, B.
13th. Coal C.
14th. Coal D, E, F.
15th. Coal G and above.

16th. Coal of unknown horizon.

17th. For species of plants also found in Europe.

As the list of species following the table indicates the reference of each locality to its horizon, marked by the number of the column, the student will find here, to my belief, all the data which it may be interesting to consult for a full acquaintance with the Coal Flora and its distribution. It is therefore useless to enter into detail on the progress of the vegetation from the Devonian to the Permian, on the degree of predominence of certain species either during the whole period of the vegetation or at its different stages, etc., all matters which are exposed to view in the table or which have been already considered in the second volume of this work.

A LIST

OF THE

SPECIES OF FOSSIL PLANTS

FOUND AT EACH LOCALITY.

---

1. MIDDLE DEVONIAN, (*VII of the Penna. Reports.*)

a. New York State, (*List of Dawson.**)

Calamites inornatus, Cordaites Robbii (?) C. angustifolius, Bornia radiata, Cyclopteris incerta, Didymophyllum reniforme, Lepidodendron Gaspianum, L. corrugatum, Psilophyton princeps, Rhachiopteris pinnata, R. striata, R. tenuistriata

b. Pennsylvania.

Huntingdon (*in Penna. Rept., 1848, by Rogers.*)
Lepidodendron primæve.

2. CHEMUNG GROUP. (VIII.)

a. New York State. (*List of Dawson, l. c.*)

Acanthophytum spinosum, Archæopteris Halliana, A. Jacksoni, Caulopteris Lockwoodi, Lepidodendron chemungense, L. corrugatum, Psilophyton princeps, Ptilophytum Vanuxemi, Rhachiopteris striata, Sigillaria Vanuxemi, Stigmaria exigua, Syringodendron gracile.

b. Pennsylvania.

Lepidodendron chemungense.

---

* The list is copied from the 16th Annual Report of New York State by Hall, p. 99.

3. CATSKILL GROUP. (IX.)

a. NEW YORK STATE. (*List of Dawson, l. c.*)

Archæopteris (cyclopteris) Jacksoni, Lepidodendron Gaspianum, Psilophyton princeps, Rhachiopteris punctata, R. cyclopteroides, Sigillaria simplicitas.

b. MAINE. (*List of Dawson, l. c.*)

Archæopteris Browni, A. Jacksoni, Lepidodendron Gaspianum, Lepidostrobus globosus, L. Richardsoni, Leptophlœum rhombicum, Psilophytum princeps, Sphenopteris Hitchcockiana.

c. PENNSYLVANIA.

Meshoppen.
Archæopteris hybernica, A. macilenta, A. minor, A. obtusa, A. Rogersi, A. sphenophylloides.

Coxton Narrows.
Archæopteris Hybernica, A. Rogersi, A. minor, Dendrophycus Desori, Rhacophyllum truncatum.

Montrose, Susquehanna.
Archæopteris obtusa.

Tioga County.
Archæopteris minor, Dictyophytum ramosum.

---

4. POCONO SANDSTONE. (X.)

a. PENNSYLVANIA.

Mauch Chunk.
Archæopteris Bockschiana, A. obtusa, A. minor.

Pottsville.
Archæopteris Bockschiana, A. obliqua.

Sideling Hill, Huntingdon Co.
Knorria acicularis, Lepidodendron species, Rhachiopteris fragments, Sphenopteris flaccida, Stigmaria minuta, Stigmatocanna (?) Ulodendron species.

c. West Virginia.

New River.

Alethopteris Helenæ, Archæopteris Halliana, Megalop teris Dawsoni, Pseudopecopteris obtusiloba, Spheno phyllum antiquum.

Lewis Tunnel.

Archæopteris Alleghaniensis, A. Rogersi, A. Bockschiana, Lepidodendron corrugatum, Pseudopecopteris Virginiana, Triphyllopteris Lescuriana.

---

5. SUB-CONGLOMERATE. (XI.)

a. Pennsylvania.

Mauch Chunk.

Sphenopteris furcata.

Pottsville.

Dendrophycus Desorii, Lepidocystis fraxiniformis, Lepi dodendron corrugatum, Stigmaria minuta.

b. West Virginia.

Conglomerate Series.

Alethopteris grandifolia, Alethopteris Helenæ, Archæop teris Halliana, Bornia radiata, Calamites approximatus, C. cannæformis, C. Roemeri, (?) Callipteridium rugosum, Cordaites Robbii, Lepidodendron dichotomum, Megalopteris Hartii, M. Sewelleni, Neuropteris Smithii, N. tenuifolia, Odontopteris gracillima, O. Newberryi, Pseudopecopteris obtusiloba, Sphenophyllum antiquum, Sphenopteris adi antoides, S. Hoeninghausii, S. spinosa.*

c. Ohio.

Newark— Waverly Sandstone.

Bornia radiata, Lepidodendron corrugatum, Ptilophyton lineare.

Rushville— Waverly Sandstone.

Alethopteris Bunburyi, A. maxima, Annularia cuspidata,

* See remarks about this list pp. 669–671.

Archæopteris denticulata, A. serrata, A. stricta, Asterophyllites erectifolius, var. of A. equisetiformis, (?) A. gracilis, Cardiocarpus Newberyi, Cordaites communis, Lepidodendron clypeatum, L. Gaspianum, L. quadrilaterale, L. Rushvillense, Megalopteris dentata, M. Hartii, M. lata, M. minima, M. ovata, M. serrata, Orthogoniopteris clara, O. Gilberti, Protoblechnum Holdeni, Ptilophyton lineare, P. Vanuxemi, Sphenopteris Ballantini, S. marginata, Sphenophyllum tenerrimum, Trigonocarpus Schultzianus, var. minor.

d. ILLINOIS.

Chester Group.

Alethopteris Helenæ, Asterophyllites equisetiformis, var. Conostychus ornatus, Halonia tuberculata, Knorria imbricata, Lepidocytis bullatus, Lepidodendron costatum, L. Sternbergii, L. turbinatum, L. Veltheimianum, Lepidophloios laricinus, Lepidostrobus species, Lesleya grandis, Megalopteris fasciculata, Neuropteris capitata, N. Germari, N. tenuifolia, Pseudopecopteris anceps, Rhacophyllum flabellatum, Sigillaria hexagona, (?) Sphenopteris cristata, S. tridaclytites.

Port Byron.

Alethopteris species, Cardiocarpus bicuspidatus, Conostychus ornatus, Cordaites diversifolius, Eremopteris elegans, Knorria princeps, (?) Lepidodendron crenatum, L. Rushvillense, L. Volkmannianum, Megalopteris abreviata, M. fasciculata, M. Southwellii, Pseudopecopteris Pluckneti, Sphenopteris mixta (?).

e. INDIANA

Whetstone Beds.

Lepidodendron Veltheimianum, L. Rushvillense, Lepidophloios crassicaulis, Neuropteris biformis, N. Elrodi, N. Smithii, Pseudopecopteris muricata, Sphenopteris Hoeninghausii, S. tridactylites, Ulodendron minus.

f. GEORGIA.

Dade County.

Alethopteris lonchitica ?, Asterophyllites gracilis, Calamites cruciatus, Calamostachys lanceolata, Cardiocarpus ovalis, C. pusillus, C. longicollis, Dictyopteris rubella, Equisetites gracilis, Eremopteris microphylla, Lepidodendron rigidum, L. Sternbergii, L. Veltheimianum, L. vestitum, Neuropteris Elrodi, N. Smithii, Pecopteris Georgiana, Pseudopecopteris decipiens, P. muricata, P. nervosa, P. speciosa, Sphenopteris divaricata, S. Gravenhorstii, var. S. Hoeninghausii, S. Larischii, S. communis, Volkmannia crassa.

g. TENNESSEE.

Tracy and Lower Sewanee.

Alethopteris Evansii, A. grandifolia, A. Helenæ, A. Serlii, Asterophyllites grandis, Bornia radiata, Cardiocarpus ellipticus, C. Harveyi, C. latior, C. longicollis, C. minus, Danæites macrophyllus, Eremopteris divaricata, E. elegans, Lepidodendron clypeatum, L. longifolium, L. marginatum, L. rigidum, L. Rushvillense, L. Sternbergii, L. Veltheimianum, Lepidophloios laricinus, Lepidostrobus Goldenbergii, var. Neuropteris biformis, N. Elrodi, N. Smithii, N. subfalcata, N. tenuifolia, Odontopteris Newberryi, O. sphenopteroides, Pecopteris acuta, P. serrula, P. serrulata, Pseudopecopteris muricata, P. latifolia, Rhabdocarpus multistriatus, Rhacophyllum flabellatum, R. lactuca, et var. angustiloba, R. truncatum, Sigillaria Lescurii, Sphenopteris crenata, S. elegans, S. Hoeninghausii, S. linearis, S. communis, S. furcata, S. mixta, S. pseudo-Murrayana, S. Tracyana, S. tridactylites, Trigonocarpus ampullæformis, T. Schultzianus, var. minor, Ulodendron minus, Whittleseya integrifolia.

Rockwood.

Eremopteris Cheathami, Neuropteris biformis, Rhabdocarpus Bockschianus, Sphenopteris Royi, Trigonocarpus ampulæformis, T. olivæformis.

h. ALABAMA

Helena, Shelby County.

Alethopteris Helenæ, Asterophyllites gracilis, Calamites ramosus, Eremopteris artemisiæfolia, E. crenulata, E. dissecta, E. flexuosa, E. microphylla, Lepidodendron clypeatum, L. squamiferum, L. Veltheimianum, Oligocarpia

Alabamensis, Pecopteris angustissima,? P. Hallii, Pseudopecopteris decipiens, P. latifolia, P. nervosa, P. polyphylla, P. speciosa, P. trifoliata, Rhabdocarpus clavatus, Sigillaria Dournaisii, Sphenopteris elegans, S. Hoeninghausii, S. Larischii, S. microcarpa, S. quercifolia, S. trichomanoides, Ulodendron commutatum.

Montevallo Mines.

Lepidostrobus Aldrichi, Lepidodendron Veltheimianum, Lepidophloios crassicaulis, Neuropteris Elrodi, N. inflata, N. biformis, Ulodendron Majus, U. Minus.

Black Lick Seam, Jefferson County.

Callipteridium Aldrichi, Neuropteris Smithii, Pseudopecopteris macilenta, P. muricata, P. nervosa, Lepidodendron Sternbergii, Rhabdocarpus clavatus.

Warrior Coal Seam.

Annularia Dawsoni, Asterophyllites equisetiformis, Calamites Suckovii.

Alabama. Divers Localities not indicated.

Asterophyllites foliosus, Bornia radiata, Calamites approximatus, C. cannæformis, C. cistii, C. Suckovii, Cordaites borassifolius, Lepidodendron aculeatum, Lepidodendron dichotomum, L. marginatum, Lepidophloios laricinus, Neuropteris biformis, N. reniformis, Pseudopecopteris latifolia, Sigillaria monostigma, S. reticulata, Tæniopteris Smithii, Trigonocarpus Noeggerathi, T. Parkinsoni.

i. Arkansas.

Various Coal Beds of the same horizon.

Alethopteris Helenæ, Annularia radiata, Asterophyllites gracilis, Bornia radiata, Calamites ramifer, C. ramosus, Calamostachys lanceolata, Callipteridium Aldrichi, C. Owenii, Cardiocarpus affinis, C. annulatus, C. bicuspidatus, C. circularis, C. conglobatus, C. crassus, C. dilatatus, C. diplotesta, C. Harveyi, C. ingens, C. late-alatus, C. longicollis, C. speciosus, C. mamillatus, C. marginatus, C. orbiculatus, C. ovalis, C. patens, C. pachytesta, C. pusillus, C. simplex, Carpolithes cerasiformis, C. granularis, C. perpusillus, C.

transectus, Conostychus Broadheadi, Cordaicarpus Gutbieri, C. lineatus, C. ovatus, C. stabilis, Cordaites diversifolius, Dictyopteris obliqua, Eremopteris artemisiæfolia, E. flexuosa, E. crenulata, Halonia pulchella, Knorria imbricata, Lepidocystis obtusus, L. quadrangulatus, Lepidodendron corrugatum, L. diplotegeinides, L. modulatum, L. Rushvillense, L. rhombicum, L. Veltheimianum, L. vestitum, L. Volkmannianum, Lepidophyllum acuminatum, L. Campbellianum, L. majus, L. brevifolium, Lycopodites arborescens, L. flexifolius, L. Meekii, L. pendulus, L. simplex, Macrostachya infundibuliformis, Megalopteris rectinervis, Neuropteris fimbriata, N. hirsuta, N. Loschii, N. Smithii, N. tenuifolia, Odontopteris Brardii, O. neuropteroides, Pecopteris Georgiana, P. erosa, P. serrulata, Pseudopecopteris cordato-ovata, P. decipiens, P. irregularis, P. latifolia, P. macilenta, P. muricata, P. nervosa, P. nummularia, Pinnularia capillacea, Rhabdocarpus amygdalaeformis, R. Beinertianus, R. clavatus, R. Jacksonianus, R. latemarginatus, R. minutus, R. multistriatus, Rhacophyllum adnascens, R. filiciforme, Sigillaria, Brardii, S. Lescurii, S. reticulata, Sorocladus stellatus, Sphenopteris chærophylloides, S. communis, S. crenata, S. flexicaulis, S. furcata, S. Harveyi, S. Hoeninghausii, S. inequilateralis, S. mixta, S. obovata, S. pseudo-Murrayana, S. tridactylites var., Sphenophyllum bifurcatum, S. longifolium, S. latifolium var., S. saxifragæfolium, S. Schlotheimii, Stigmaria ficoides, Trigonocarpus ampullæformis, T. Parkinsoni, T. subcylindricus, T. olivæformis, Whittleseya microphylla.

---

6. INTER-CONGLOMERATE XII

a. Pennsylvania

*Campbell's Ledge.**

Alethopteris Davreusii, A. grandis, A. pteroides, A. Serlii, Annularia longifolia, A. sphenophylloides, Asterophyllites equisetiformis, A. foliosus, A. longifolius, A.

*The localities for species found at and around Pittston, Pa., have been kindly furnished by Mr. R. D. Lacoe.

rigidus, Bornia radiata, Calamites ramifer, C. ramosus, Callipteridium Grandini, Cardiocarpus annularis, C. annulatus, C. apiculatus, C. congruens, C. crassus, C. dilatatus, C. diminutivus, C. diplotesta, C. ellipticus, C. elongatus, C. fasciculatus, C. late-alatus, C. latus, C. minus, C. orbicularis, C. pachytesta, C. pussillus, C. regularis, C. simplex, C. zonulatus, Carpolithes acuminatus, C. cistula, C. elongatus, C. latior, C. perpusillus, C. retusus, Cordaianthus baccifer, C. ovalis var., C. flexuosus, C. (gemmifer) ovatus, C. rugosus, C. scaber, C. spicatus, Cordaicarpus stabilis, Cordaites borassifolius, C. grandifolius, C. Lacoei, Lepidocystis angularis, L. bullatus, L. fraxiniformis, L. vesicularis, Lepidodendron clypeatum, L. diplotegioides,? L. modulatum, L. Sternbergii, L. Veltheimianum, Lepidophyllum acuminatum, L. Campbellanium, L. cultriforme, L. gracile, L. lanceolatum, Lepidostrobus connivens, L. hastatus, L. incertus, L. variabilis, Lycopodites simplex, Neuropteris aspera, Pecopteris dentata, P. elliptica, P. longifolia, P. notata, Pseudopecopteris glandulosa, P. irregularis, P. latifolia, P. Loschii, P. muricata, P. nervosa, P. nummularia, P. Sillimanni, Rhabdocarpus amygdalæformis, R. Danai, R. Jacksonianus, R. late-costatus, R. mamillatus, R. minutus, R. multistriatus, Rhacophyllum Gutbierianum, Sphenopteris furcata, Sphenophyllum emarginatum, S. Schlotheimii, Spirangium appendiculatum, S. intermedium, S. multiplicatum, Sporocystis planus, Tæniopteris truncata, Trigonocarpus juglans, Ulodendron majus.

West Pittston, Brewery Cut.

Annularia tuberculata, Lepidocystis obtusus, Lepidophyllum cultriforme, Macrostachya lanceolata.

Slippery Rock Creek, Lawrence County.

Taonurus marginatus.

Venango County, Near Oil City.

Halonia tuberculata.

b. OHIO.

Youngstown.

Cardiocarpus annulatus, C. elongatus, Carpolithes fragarioides, C. retusus, Danæites macrophyllus, Lepidostrobus mirabilis, Neriopteris lanceolata, Odontopteris gracillima, O. Newberryi, Rhabdocarpus acuminatus, R. late costatus, Sigillaria pulchra.

Cuyahoga Falls, Summit County.

Cardiocarpus bicuspidatus, C. latus, C. minus, C. orbicularis, Odontopteris Newberryi, Rhabdocarpus carinatus, R. Jacksonianus, R. lævis, Sigillaria acuminata, S. elliptica, Trigonocarpus juglans, T. multicarinatus, T. ornatus, T. trilocularis, Whittleseya elegans.

Talmadge.

Alethopteris grandifolia, Cardiocarpus bicuspidatus, C. samaræformis, Carpolithes retusus, Lepidostrobus mirabilis, Neriopteris lanceolata, Rhabdocarpus apiculatus, R. carinatus, Trigonocarpus tricuspidatus.

Jackson's Shaft, Perry County.

Alethopteris Helenæ, Lepidodendron Veltheimianum, Lepidophloios laricinus.

LOCALITIES UNDETERMINED.

INTER-CONGLOMERATE OR COAL A.

Tennessee.

Main Sewanee and Ætna Veins, Base of the Coal Measures.

Alethopteris lonchitica, Calamites cruciatus, Lepidodendron aculeatum, L. elegans, L modulatum, L. obovatum, L. rimosum, L. dichotomum, L. vestitum, Sigillaria attenuata, S. Brardii, S. mamillaris, S. Menardi, S. tessellata, Sphenopteris linearis, Stigmaria ficoides, Syringodendron cyclostygma, S. pachyderma, Trigonocarpus Noeggerathi, T. olivæformis, T. Parkinsoni, T. Saffordi, Ulodendron minus.

ANTHRACITE BASINS.

7. MIDDLE OR PRODUCTIVE COAL MEASURES.

PENNSYLVANIA.

COAL A.

Shamokin.

Asterophyllites equisetiformis, Annularia inflata, A. longifolia et var. angustifolia, A. sphenophylloides, Cordaicarpus ovatus, Knorria acicularis, Lepidocystis quadrangularis, Lepidodendron obovatum, L. Sternbergii, Lepidophyllum lanceolatum, Neuropteris angustifolia, N. hirsuta, N. rotundifolia, Sigillaria monostigma, Sphenophyllum Schlotheimii, Sphenopteris trichomanoides, Ulodendron punctatum.

Lehigh Summit.

Lepidodendron aculeatum, L. Mielicki, L. vestitum, Pseudopecopteris Newberryi, Sigillaria lepidodendrifolia, S. discoidea.

Trevorton, (Low Coal.)

Cardiocarpus marginatus, C. plicatus, C. Trevortoni, Carpolithes acuminatus, Dictyopteris obliqua, Rhabdocarpus amygdalæformis, Sigillaria Brongniarti, S. Cortei, S. Lescurii, S. obovata, Syringodendron cyclostigma, S. pachyderma.

Coxton & Everhart Creek.

Alethopteris Serlii, Cordaicarpus costatus, Lepidophyllum acuminatum, L. oblongifolium, Sigillaria monostigma, Trigonocarpus Bertholetiformis, T. Starkianus, Ulodendron ellipticum.

Johnstown, Cambria Co., (Low Coal.)

Lepidophyllum acuminatum, L. brevifolium, L. obtusum.

Gaines, Tioga County.

Lepidodendron clypeatum, L. rimosum.

Broad Top, Huntingdon County.

Lepidodendron Tijoui.

COAL B.

Archbald.

Asterophyllites equisetiformis, Lepidophyllum campbellianum, L. minutum, Lepidostrobus ovatifolius, Macrostachya communis, Neuropteris Desorii, N. inflata, Pecopteris pennæformis, Volkmannia fertilis.

COAL B OR C.

Carbondale.

Alethopteris Serlii, Calamites cannæformis, C. cistii, C. Suckowii, Lepidodendron aculeatum, L. carinatum, L. dichotomum, L. distans, L. modulatum, L. obtusum, L. rimosum, L. Veltheimianum, L. vestitum, Lepidophloios crassicaulis, Lepidophyllum lanceolatum, Sigillaria dilatata, S. polita, S. stellata, S. tessellata var. elegans, Stemmatopteris gigantea.

COAL C.

Archbald.

Lepidodendron aculeatum, L. clypeatum, L. modulatum, L. obovatum, L. vestitum.

Shamokin.

Callipteridium Sullivantii, Cordaianthus rugosus, Pseudopecopteris nervosa, Rhabdocarpus multistriatus.

Boston mine, Pittston.

Cardiocarpus divergens, C. pachytesta, Carpolithes granularis, Cordaianthus ebracteatus, Cordaites borassifolius, C. lingulatus, C. validus, Knorria imbricata, Lepidodendron clypeatum, L. dichotomum, L. Veltheimianum, L. vestitum, Lepidophyllum linearifolium, Lepidostrobus prælongus, Odontopteris Reichiana, Sigillaria tessellata, Sphenophyllum Schlotheimii, Trigonocarpus Dawesii.

Ontario Colliery, Pittston.

Lepidodendron aculeatum, L. clypeatum, L. dichotomum, L. modulatum, L. obovatum, L. rimosum, L. scutatum, L. Sternbergii, L. Veltheimianum, L. Volkmannianum, Neuropteris angustifolia, Pecopteris pennæformis, Rhabdocarpus mamillatus, R. multistriatus, Sigillaria Brardii, S. Menardi, S. Yardleyi, Volkmannia prælonga.

"Butler Dam," Pittston.

Annularia longifolia, Caulopteris Cistii, Dictyopteris obliqua, Lepidodendron dichotomum, Lepidostrobus ovatifolius, Pecopteris vestita, Rhabdocarpus subglobosus, Rhacophyllum filiciforme, Trigonocarpus Starkianus.

COAL C?.

Hughestown Deep Shaft, Pittston.

Carpolithes perpusillus, Lepidophyllum lanceolatum, Neuropteris coriacea, Odontopteris Schlotheimii, Pecopteris notata, P. pennæformis.

COAL B or C.

Carbon Hill Shaft, Old Forge.

Odontopteris abbreviata, Rhabdocarpus multistriatus.

COAL D.

Carbon Hill Tunnel, Old Forge.

Alethopteris Serlii, Asterophyllites equisetiformis, Callipteridium Sullivantii, Lepidodendron obovatum, L. Sternbergii, Lepidophyllum majus, Neuropteris hirsuta, Pseudopecopteris anceps, P. Plucknetí, Rhabdocarpus multistriatus, Sigillaria mamillaris, Sphenopteris dissecta, Trigonocarpus juglans.

Tompkins Mine, Pittston.

Lepidodendron obovatum, L. modulatum, L. Veltheimianum, Lepidophyllum striatum, Neuropteris aspera, N. Clarksoni, Pecopteris vestita, Pseudopecopteris anceps.

Pittston.

Sphenopteris tridactylites.

COAL C OR D.

Wilkes Barre.

Lepidophyllum affine, L. brevifolium, L. linearifolium, Lepidostrobus ornatus, Pseudopecopteris nervosa.

Connell's Mine, Minooka.

Alethopteris Serlii, Sphenophyllum Schlotheimii

Taylorville.

Alethopteris Serlii, Pecopteris erosa, Pseudopecopteris

nervosa, var. Sauveurii, Rhabdocarpus multistriatus, Trigonocarpus Schultzianus.

Susq. Anthr. Coal Co. Mine, Moosic coal C.

Neuropteris Carrii.

Brown Colliery, Pittston, coal D or E.

Annularia inflata, A. longifolia, var. angustifolia, Asterophyllites rigidus, Calamites ramosus, C. Suckowii, Lepidodendron clypeatum, L. diplotegioides, L. latifolium, L. longifolium, L. modulatum, L. obovatum, L. rugosum, L. Veltheimianum, Lepidophyllum elegans, L. lanceolatum, L. lancifolium, L. striatum, Lepidophloios auriculatus, Lepidostrobus Butleri, L. lanceolatus, Neuropteris angustifolia, N. callosa, N. capitata, N. fimbriata, N. heterophylla, N. Loschii, N. plicata, N. trichomanoides, Odontopteris Alpina, Oligocarpia Gutbieri, Pecopteris abbreviata, P. arborescens, P. Candolliana, P. dentata, P. oreopteridis, P. pennæformis, P. unita, P. villosa, Pseudopecopteris anceps, P. Pluckneti, Sigillaria Brardii, Sphenophyllum cornutum, S. Schlotheimii, Sphenopteris Gravenhorstii, Ulodendron elongatum, U. majus, U. minus.

COAL E.

Rausch Gap.

Neuropteris Grangeri, Pecopteris Bucklandi, P. Cistii, P. pennæformis, Sigillaria Lorenzii, S. Yardleyi, Stigmaria amœna.

Pottsville (Mammoth Vein.)

Alethopteris lonchitica, Lepidocystis quadrangularis, Lepidostrobus lanceolatus, Neuropteris angustifolia, N. cordata, Stigmaria umbonata.

Butler Mine, Pittston.

Alethopteris Serlii, Annularia longifolia var. angustifolia, Calamites approximatus, C. cannæformis, C. Cistii, C. dubius, C. ramosus, C. Suckowii, Cardiocarpus annulatus, Carpolithes Butlerianus, Caulopteris obtecta, Cordaites borassifolius, C. diversifolius, C. Lacoei, C. validus, Knorria imbricata, Lepidocystis pectinatus, Lepidodendron aculeatum, L. clypeatum, L. modulatum, L. obovatum, L. obtusum, L. rimosum, L. Veltheimianum, Lepidostrobus affinis, L. Butleri, L. variabilis, Lepidophyllum acuminatum, L. majus, L. striatum, Neuropteris fimbriata, N. plicata, Pecopteris abbreviata, P. acuta, P. Candolliana, P. arguta, P. Cistii, P. dentata, P. elliptica, P. Miltoni, R. oreopteridis, P. pennæformis, P. pteroides, P. villosa, Pseudopecopteris nervosa, Rhabdocarpus clavatus, R. Jacksonianus, R. multistriatus, Rhacophyllum filiciforme, R. lactuca, Sphenophyllum Schlotheimi, Sigillaria Brochanti, S. Serlii, Trigonocarpus Dawesii, T. juglans, Ulodendron majus, U. punctatum.

Penna. Coal Co. (Shaft No. 6), Pittston.

Calamites approximatus, C. Cistii, C. cruciatus, C. dubius, Lepidodendron aculeatum, L. modulatum, L. Veltheimianum, Lepidophyllum rostellatum, L. striatum Pecopteris oreopteridis, Sigillaria mamillaris.

Yatesville (Railroad Cut).

Alethopteris ambigua, A. Pennsylvanica, A. Serlii, Annularia inflata, Asterophyllites equisetiformis, Caulopteris obtecta, Lepidodendron rimosum, Lepidostrobus ovatifolius, Lepidophyllum affine, L. lanceolatum, Neuropteris angustifolia, N. Collinsii, N. hirsuta, N. inflata, N. rarinervis, Rhabdocarpus multistriatus, Sphenophyllum Schlotheimii.

COAL E OR ABOVE.

Oakwood Colliery, Wilkes-Barre.

Alethopteris Serlii, Annularia longifolia, Calamites Suckowii, Callipteridium rugosum, Cardiocarpus ellipticus, C. fasciculatus, Cordaites diversifolius, C. validus, Lepidodendron aculeatum, L. Veltheimianum, Lepidostrobus affinis, Neuropteris angustifolia, N. cordata, N. fimbriata, N. Germari, N. hirsuta, N. inflata, N. Loschii, N. plicata, N. rarinervis, N. rotundifolia, Odontopteris obtusa, O. Schlotheimii, Pecopteris Candolliana, P. dentata, P. Miltoni, P. oreopteridis, P. vestita, Pseudopecopteris decipiens, P. irregularis, Rhacophyllum trichoideum, Sigillaria lævigata.

COAL E OR F.

Port Griffith Railroad Cut, below Pittston.

Alethopteris Serlii, Annularia inflata, A. longifolia, A. sphenophylloides, Asterophyllites equisetiformis, Calamites cannæformis, Dictyopteris obliqua, D. Scheuchzeri, Halonia tortuosa, Lepidocystis angularis, L. quadrangularis, Lepidodendron aculeatum var. caudatum, L. obovatum, Lepidophyllum lanceolatum, L. striatum, Lepidophloios laricinus, Lepidostrobus ovatifolius, Neuropteris Loschii, N. plicata, N. vermicularis, Pecopteris Miltoni, P. villosa, Pseudopecopteris denudata, Rhabdocarpus Jacksonianus, Rhacophyllum Gutbierianum, Sigillarioides radicans, Sphenophyllum emarginatum, S. longifolium, S. Schlotheimii, Sphenopteris chærophylloides, S. furcata, S. tridactylites, Trigonocarpus Adamsii.

Port Griffith, Switch-Back.

Carpolithes perpusillus, Caulopteris Cistii, C. Locoei, Neuropteris Griffithii, Pecopteris arguta, P. emarginata, P. longifolia, P. robusta, P. unita, Sigillaria Brardii, S. lævigata, S. Menardi, S. Serlii, Sphenophyllum Schlotheimii, Stemmatopteris punctata.

COAL E & F.

Stanton and Empire Mines, Wilkes-Barre.

Alethopteris lonchiticá, Calamites approximatus, C. ramosus, Callipteridium Massilloneum, C. Oweni, Cardiocarpus minus, C. pusillus, Carpolithes minimus, Caulopteris Cistii, Dictyopteris obliqua, Lepidocystis angularis, Lepidodendron Mielickii, Lepidophyllum acuminatum, L. lanceolatum, L. majus, L. Mansfieldi, L. Stantoni, L. striatum, Neuropteris Carrii, (?) N. Clarksoni, N. coriacea, N. Desorii, N. fimbriata, N. plicata, N. rarinervis, N. rotundifolia, N. tenuifolia, and var. angustifolia, N. vermicularis, Pecopteris Bucklandi, P. elliptica, P. Miltoni, P. unita, Pseudopecopteris anceps, P. cordato-ovata, P. decurrens, P. latifolia, P. nervosa, P. Pluckneti, Rhabdocarpus apiculatus, R. carinatus, R. Jacksonianus, R. late-costatus, R. multistriatus, Rhacophyllum filiciforme, R Gutbierianum, R.

lactuca, Sigillaria tessellata, Sphenophyllum cornutum, S. longifolium, Trigonocarpus Adamsii, T. Dawesii, T. Schultzianus.

COAL E, (?) F, (?) G. (?)

Wilkes-Barre.

Alethopteris Serlii, Asterophyllites longifolius, Calamites Cistii, Cordaianthus rugosus, Dictyopteris obliqua, Lepidodendron rhombicum, L. vestitum, Lepidophyllum tumidum, Lepidostrobus ornatus, Neuropteris callosa, N. Clarksoni, N. crenulata, N. Desorii, N. Grangeri. Pecopteris Cistii, P. dentata, P. oreopteridis, P. unita, Pseudopecopteris anceps, P. cordato-ovata, P. nervosa, P. Newberryi, P. Pluckneti, Sigillaria Lescurii, S. Menardi, S. pachyderma, S. reniformis, S. rugosa, S. Sillimanni, S. tessellata, Ulodendron punctatum.

COAL E OR F.

Kingston.

Lepidophyllum lanceolatum, Neuropteris Carrii, N. reniformis.

COAL F.

Plymouth.

Calamites approximatus, Lepidodendron cuspidatum, L. obovatum, Sigillaria Brardii, S. cuspidata, S. elliptica, S. Lacoei, S. lævigata, S. leptoderma, S. Lescurii, S. mamillaris, S. marginata, S. Menardi, S. obliqua, S. ovalis, S. Pittstoniana, S. pyriformis, S. reniformis, S. Saullii, S. Sillimanni, S. tessellata, S. venosa, S. Voltzii, S. Williamsii, S. Yardleyi, Syringodendron cyclostigma, S. pachyderma.

F OR NEXT ABOVE F.

Orchard Mine Vein.

Annularia longifolia var. angustifolia, A. radiata, Asterophyllites anthracinus, Calamites dubius, Caulopteris Mansfieldi, Lepidodendron Brittsii, Lycopodites Lacoei, Neuropteris callosa, N. crenulata, N. heterophylla ? N. plicata, Pecopteris Candolliana, P. Clintoni, P. dentata, P. elliptica, P. oreopteridis, P. pennæformis, Pseudopecopteris anceps, P. dimorpha, P. nummularia, P. Pluckneti, Rhacophyllum filiciforme, Sigillaria Brardii, S. mamillaris, S. Menardi, S. obliqua, S. venosa, Sphenophyllum Schlotheimi.

COAL F.

Pittston.

Alethopteris grandis, Cardiocarpus divergens, Cordaicarpus costatus, C. cinctus.

Maltby.

Alethopteris lonchitica, A. Pennsylvanica, A. Serlii, Callipteridium Sullivantii, Pecopteris lanceolata, P. longifolia, P. villosa, Rhabdocarpus Bockschianus, Sigillaria orbicularis.

Seneca Mine, Pittston.

Cordiacarpus Mansfieldi, Knorria imbricata, Lepidodendron Veltheimianum, L. Volkmannianum, Lepidophloios laricinus, Rhabdocarpus acuminatus, R. insignis, Sigillaria Brardii, S. elliptica, S. Menardi, S. orbicularis, Ulodendron minus.

Tompkins Mine, Pittston.

Cardiocarpus ellipticus, Dictyopteris obliqua, Trigonocarpus Starkianus, Ulodendron punctatum.

COAL G.

Olyphant.

Annularia longifolia and var. angustifolia, Asterophyllites equisetiformis, Calamites cannæformis, C. cruciatus, C. dubius, C. Suckowii, Callipteridium rigidum, C. rugosum, C. Sullivantii, Carpolithes perpusillus, Caulopteris Cistii, C. Lacoei, C. Mansfieldi, Eremopteris artemisiæfolia, Lepidodendron aculeatum, L. clypeatum, L. latifolium, L. modulatum, Lepidophloios laricinus, Lepidophyllum acuminatum, L. majus, L. Mansfieldi, L. rostellatum, Lepidostrobus Goldenbergii, L. prælongus, L. variabilis, Lycopodites Lacoei, Macrostachya communis, Megaphytum Grand'Euryi, Neuropteris Clarksoni, N. Collinsii, N. coriacea, N. crenulata, N. fimbriata, N. hirsuta, N. inflata, N. Loschii, N. rarinervis, N. vermicularis, Odontopteris Alpina, O. Worthenii, Pecopteris acuta, P. arborescens, P. arguta, P. Cistii, P. dentata, P. elegans, P. elliptica, P. oreopteridis, P. pennæformis, P. polymorpha, P. pteroides, P. quadratifolia, P. robusta, P. unita, P. vestita, P. villosa, Pseudopecopteris cordato-ovata, P. dimorpha, P. latifolia, P. nervosa, P. Newberryi, P. Pluckneti, Rhabdocarpus Jacksonianus, R. multistriatus, Rhacophyllum Clarkii, R. obtusilobum, R. expansum, R. filiciforme, R. Guthbierianum, R. lactuca, Sigillaria cyclostigma, S. Lacoei, S. Lescurii, S. mamillaris, S. obliqua, S. pyriformis, S. tessellata, S. Voltzii, S. Williamsii, Sphenopteris fascicularis, S. tenella, S. tenuifolia, Sphenophyllum oblongifolium, S. Schlotheimii, Stemmatopteris cyclostigma, S. hirsuta, S. peltigera, S. punctata, S. squamosa, Ulodendron punctatum.

COAL G.

Enterprise Colliery, Plainsville.

Annularia sphenophylloides, Callipteridium Massilloneum, Pecopteris Clarkii, P. pteroides, Pseudopecopteris anceps

COAL M.

Gate and Salem Vein, Pottsville.

Annularia minuta, A. sphenophylloides, Asterophyllites equisetiformis, Calamites cannæformis, C. disjunctus, C. ramosus, C. Suckowii, Callipteridium rugosum, Cardiocarpus bicuspidatus, Dictyopteris obliqua, Neuropteris angustifolia, N. callosa, N. cordata, N. crenulata, N. dentata, N. Desorii, N. fimbriata, N. fissa, N. Germari, N. gibbosa, N. Grangeri, N. hirsuta, N. plicata, N. rotundifolia, N. tenuifolia, N. trichomanoides, Odontopteris alata, O. Schlotheimii, O. tenuinervis, O. subcrenulata, Pecopteris arborescens, P. concinna, P. distans, P. elegans, P. abbreviata, P. Miltoni, P. nodosa, P. notata, P. polymorpha, P. unita, Pseudopecopteris decurrens, P. incompleta, P. Loschii, P. pussilla, P. Sheaferi, Rhacophyllum fimbriatum, R. hirsutum, Sphenophyllum marginatum, S. filiculme.

South Salem Vein, Pottsville.

Lescuropteris adiantites, Oligocarpia flagellaris, Pecopteris arguta, P. oreopteridis, P. platyrachis, P. pennæformis, Pseudopecopteris latifolia, P. muricata.

Gate Vein, New Philadelphia.

Alethopteris marginata, Carpolithes bifidus, Lepidophyllum affine, Neuropteris callosa, N. fimbriata, N. Grangeri, N. Desorii, Pecopteris arborescens, P. arguta, P. oreopteridis, P. platyrachis, P. pennæformis, Rhacophyllum lactuca, Sigillaria obliqua, S. reniformis, Sphenophyllum filiculme, Stemmatopteris punctata.

Salem Vein, Port Carbon.

Cordaites diversifolius, Neuropteris dentata, N. Grangeri, N. Rogersi, Pecopteris arborescens, P. cyathea, Sphenopteris paupercula.

Salem Vein, Tremont.

Lepidodendron icthyolepis, Neuropteris anomala, N. crenulata, Odontopteris alata, O. Schlotheimii, O. subcrenulata, Pecopteris Miltoni.

LOCALITIES OF UNCERTAIN HORIZON.

Tremont, New Vein.

Alethopteris marginata, Odontopteris Schlotheimii, Pecopteris elliptica.

Pottsville and vicinity.

Alethopteris lonchitica, Lepidodendron rimosum, Sigillaria reniformis, S. Yardleyi, Macrostachya infundibuliformis. Pecopteris oreopteridis, Knorria imbricata. Lepidostrobus lanceolatus, Neuropteris acuminata, N. dentata, Pseudopecopteris anceps.

Rhode Island.

Newport and Mount Hope, Coal Mines.

Alethopteris aquilina, A. Serlii, Annularia calamitoides, A. inflata, A. sphenophylloides, Asterophyllites equisetiformis, A. grandis, A. rigidus, A. sublævis, Dictyopteris obliqua, Lepidodendron aculeatum, L. dichotomum, Lepidostrobus lanceolatus, Lepidophyllum fallax, L. hastatum, L. lanceolatum, L. majus, L. oblongifolius, L. Stantoni, L. tumidum, Neuropteris Agassizi, N. crenulata, N. Desorii, N. Germari, N. heterophylla, N. tenuifolia, Odontopteris alata, O. Alpina, O. Brardii, O. deformata, O. neuropteroides, O. patens, Pecopteris arborescens, P. arguta, P. aspidioides, P. Candolliana, P. Clarkii, P. dentata, P. erosa, P. Miltoni, P. oreopteridis, P. pennæformis, P. platyrachis, P. quadratifolia, P. unita, Pseudopecopteris anceps, P. dimorpha, P. muricata, P. nervosa, P. Pluckneti, P. spinulosa, Rhacophyllum affine, R. Clarkii, R. filiforme, R. filiciforme, R. fimbriatum, R. hirsutum var. affine, Sphenophyllum emarginatum, S. filiculme, S. longifolium, S. Schlotheimii, Sphenopteris chærophylloides, S. cristata, S. elegans, S. Gravenhorstii, S. Hoeninghausii, S. pseudo-Murrayana, S. tridactylites.

Muddy Creek, West of Pottsville.

Neuropteris laciniata, Odontopteris squamosa, Pecopteris distans, P. Miltoni, P. unita.

Muddy Creek Vein, between Pottsville and Tremont.

Sigillaria Brardii, S. fissa, S. Menardi, S. Schimperi.

Shamokin, Horizon Unknown.

Pseudopecopteris glandulosa, P. decipiens, this last under XII.

Ashland Gap Mine.

Sigillaria attenuata, S. Lescurii.

New Philadelphia, Old Mine West of Town.

Pecopteris serrula, Sigillaria Brardii, S. dilatata, S. leioderma, S. notata, Sphenopteris plicata.

Lorberry Junction.

Annularia inflata, A. longifolia, A. sphenophylloides, Cordaicarpus costatus, Neuropteris Clarksoni, Odontopteris subcrenulata, Rhabdocarpus Jacksonianus.

Hazleton.

Cardiocarpus mamillatus, Carpolithes bifidus.

Johnston, Upper Beds.

Neuropteris obscura, Pecopteris velutina.

Tamaqua Mines.

Neuropteris obscura, N. rarinervis.

BITUMINOUS COAL REGION.

Pennsylvania.

Coal B or C.

Cannelton.

Alethopteris ambigua, A. crassa, A. lonchitica, A. Pennsylvanica, A. robusta, A. Serlii, Annularia inflata, A. longifolia, A. sphenophylloides, Asterophyllites anthracinus, A. equisetiformis, A. foliosus, A. grandis, A. rigidus, A. sublævis, Artisia species, Calamites approximatus C. cannæformis, C. disjunctus, C. ramifer, C. ramosus, C. Suckowii, Calamostachys brevifolius, Callipteridium inæquale, C. Mansfieldi, C. Sullivantii, C. neuropteroides, Cardiocarpus apiculatus, C. circularis, C. congruens, C. diplotesta, C. ellipticus, C. latus, C. mamillatus, C. marginatus, C. pusillus, C. regularis, C. speciosus, Carpolithes cerasiformis, C. cistula, C. minimus, C. perpusillus, Caulopteris Cistii, C. Mansfieldi, C. obtecta, Cordaianthus baccifer, C. dichotomus, C. ovatus, Cordaicarpus apiculatus, C. costatus, C. cinctus, C. Gutbieri, C. lineatus, C. Mansfieldi, C. ovatus, Cordaistrobus Grand'Euryi, Cordaites borassifolius, C. costatus, C. gracilis, C. Mansfieldi, C. radiatus, C. serpens, C. validus, Desmiophyllum gracile, Dicranophyllum dichotomum, D. dimorphum, Dictyopteris obliqua, D. Scheuchzeri, Equisetites occidentalis, Eremopteris artemisiæfolia, E. elegans, Halonia Mansfieldi, Lepidocystis angularis, L. fraxiniformis, L. vesicularis, Lepidodendron aculeatum, L. clypeatum, L. dichotomum, L. modulatum, L. quadratum, L. rigens, L. rugosum, L. Sternbergii, L. vestitum, Lepidophloios crassicaulis, L. dilatatus, L. laricinus, Lepidophyllum acuminatum, L. coriaceum, L. cultriforme, L. fallax, L. linearifolium, L. Mansfieldi, Lepidostrobus foliaceus, L. Goldenbergii, L. Mansfieldi, L. ornatus, L. spectabilis, L. truncatus, L. variabilis, Macrostachya aperta, M. communis, M. infundibuliformis, Megaphytum McLayi, Neuropteris Agassizi, N. angustifolia, N. anomala, N. aspera, N. auriculata, N. capitata, N. Clarksoni, N. Collinsii, N. cordata, N. coriacea, N. crenulata, N. decipiens, N. Desorii, N. fasciculata, N. fimbriata, N. Germari, N. gibbosa, N. Grangeri, N. hirsuta, N. inflata, N. Loschii, N. reniformis, N. oblongifolia, N. plicata, N. rarinervis, N. reniformis, .N Rogersi, N. rotundifolia, N. tenuifolia, N. trichomanoides, N. verbenæfolia, N. vermicularis, Odontopteris abbreviata, O. affinis, O. æqualis, O. Brardii, O. Britannica, O. cornuta, O. dilatata, O. monstruosa, O. patens, O. Schlotheimii, O. sphenopteroides, O. subcuneata, Pachypteris gracillima, Pecopteris acuta, P. Bucklandi, P. Cistii, P. Clintoni, P. cristata, P. dentata, P. erosa, P. obsoleta, P. oreopteridis, P. pennæformis, P. polymorpha, P. quadratifolia, P. squamosa, P. unita, P. velutina, P. vestita, P. villosa, Pinnularia capillacea, Pseudopecopteris anceps, P. denudata, P. incompleta, P. latifolia, P. macilenta, P. nervosa, P. Newberryi, P. nummularia, P. Pluckneti, P. Sillimanni, C. subnervosa, P. subcrenulata, Rhachiopteris squamosa, Rhabdocarpus abnormalis, R. arcuatus, R. clavatus, R. emarginatus, R. inflatus, R. Jacksonianus, R. multistriatus, R. subglobosus, R. tenax, Rhacophyllum adnascens, R. Clarkii, R. cornutum, R. expansum, R. filiciforme var. Gutbierianum, R. lactuca, Rhizomorpha sigillariæ, Sigillaria Brardii, S. elliptica, S. Grand'Euryi, S. mamillaris, S. monostigma, S. orbicularis, S. reniformis, S. scutata, S. tessellata, S. Yardleyi, Sphenophyllum emarginatum, S. longifolium, S. oblongifolium, S. Schlotheimii, Sphenopteris chærophylloides, S. cristata var. minor, S. goniopteroides, S. hymenophylloides, S. mixta, S. subalata, S. stricta, S. tridactylites, Stemmatopteris anceps, S. angustata, S. cyclostigma, S. emarginata, S. gigantea, S. hirsuta, S. insignis, S. mimica, S. polita, S. punctata, S. Schimperi, S. squamosa,

Stigmaria ficoides, Syringodendron Brongniarti, S. cyclostigma, Tæniophyllum brevifolium, T. contextum, T. decurrens, T. deflexum, Trigonocarpus Dawesii, T. grandis, T. Parkinsoni, Ulodendron punctatum.

COAL G.

Beaver County.

Asterophycus simplex.

STATE OF OHIO.

COAL A.

Massillon.

Callipteridium Massilloneum, Carpolithes retusus, Sigillaria leioderma, S. spinulosa, S. tessellata, Trigonocarpus Menzelianus.

COAL E.

Buchtel.

Neuropteris fimbriata, N. plicata, N. rarinervis, N. reniformis, Pecopteris Schimperi, P. unita, Pseudopecopteris anceps, Rhacophyllum filiciforme, Sphenophyllum emarginatum.

COAL E.

Nelsonsville.

Calamites approximatus, C. cannæformis, C. Cistii, Lepidodendron rigens, Pecopteris elliptica, P. Miltoni, Rhabdocarpus Jacksonianus, Sigillaria tessellata, Sphenophyllum emarginatum, S. angustifolium, Sphenopteris Brittsii.

Shawnee, Perry County.

Lycopodites Ortoni.

Coshocton.

Cardiocarpus bicornutus, Sigillaria Brardii, Trigonocarpus magnus.

COAL G.

Pomeroy

Neuropteris callosa, N. crenulata, N. plicata, Pecopteris arborescens, P. oreopteridis, P. pteroides, Sigillaria Brardii, S. Menardi.

Barnesville.

Alethopteris Gibsoni, Caulopteris Cistii, Sphenophyllum longifolium.

St. Clairsville.

Annularia Dawsoni, A. Emersoni, A. longifolia, Danæites Emersoni, Dictyopteris Brongniarti? Lescuropteris Moorii, Neuropteris anomala, N. fimbriata, N. hirsuta, N. obscura, N. tenuifolia, Odontopteris Bradleyi, O. Britannica, O. Schlotheimii, Pecopteris Miltoni, P. oreopteridis, P. squamosa, Pseudopecopteris nummularia, Sphenophyllum angustifolium, S. Schlotheimii, Sphenopteris Dubuissonis.

Athens.

Pecopteris arborescens.

Marietta.

Lepidodendron carinatum, Neuropteris Loschii, Pecopteris arborescens, P. nodosa.

Of uncertain horizon.

Zanesville. Divers Localities

Alethopteris Serlii, Annularia longifolia, A. sphenophylloides, Bornia radiata, Calamites dubius, Cardiocarpus marginatus, Neuropteris Grangeri, Rhabdocarpus Danai.

Guernsey County.

Trigonocarpus Bertholetiformis

Middleburg.

Sigillaria Sillimanni.

Upper Coal, Salineville.

Dictyopteris obliqua, Neuropteris cordata, Sphenophyllum Schlotheimii, Ulodendron majus.

INDIANA.

COAL B.

Spring Creek.

Alethopteris Pennsylvanica, A. Serlii, Annularia inflata, Callipteridium Oweni, Cardiocarpus marginatus, Lepidodendron dichotomum, L. Sternbergii, Neuropteris cordata, N. hirsuta, N. Loschii, Odontopteris heterophylla, O. subcuneata, Pecopteris arborescens, P. venulosa, P. villosa, Pseudopecopteris Pluckneti, Sphenophyllum emarginatum, S. longifolium.

Vermilion County.

Alethopteris ambigua, A. aquilina, A. lonchitica, A. Serlii, Cardiocarpus circularis, Lepidophyllum brevifolium, L. ovatifolium, Odontopteris squamosa, O. subcuneata, Spirangium Prendelii.

Newport.

Sigillaria icthyolepis, S. Menardi, Trigonocarpus Dawesii.

COAL C.

Eugenie, Vermilion County.

Syringodendron Porteri, Trigonocarpus olivæformis.

COAL D or E.

Sullivan County.

Lepidophyllum acuminatum, Neuropteris angustifolia, N. Desorii, Odontopteris subcrenulata, Pecopteris quadratifolia, P. pennæformis, Pseudopecopteris decipiens, P. Pluckneti, Sphenophyllum Schlotheimii, Sphenopteris spinosa, Ulodendron majus.

KENTUCKY.

COAL A.

Crittenden County Union Mines.

Alethopteris lonchitica, Calamites Suckowii, Calamostachys ovalis, Callipteridium Owenii, Lepidodendron rimosum, Lepidophloios species, Sphenopteris mediana, S. tridactylites.

COAL B.

Union County.

Alethopteris lonchitica, A. Serlii, Callipteridium Owenii, C. Sullivantii, Cordaites borassifolius, Eremopteris artemisiæfolia, Lycopodites cavifolius, Neuropteris Clarksoni, N. hirsuta, N. reniformis, N. rarinervis, N. vermicularis, Odontopteris Schlotheimii, Pseudopecopteris latifolia, P. nervosa, P. Newberryi, P. Sillimanni, P. anceps, Sigillaria reniformis, S. attenuata, S. corrugata, S. discoidea, S. stellata, S. tessellata, Sphenopteris Hildreti, S. furcata.

Greenup and Carter Counties.

Annularia longifolia, A. tuberculata, Alethopteris Serlii, Calamites Suckowii, Dictyopteris obliqua, Lepidodendron Sternbergii, Lepidostrobus princeps, Neuropteris hirsuta, Pecopteris oreopteridis, P. unita, Pinnularia capillacea, Pseudopecopteris latifolia, Sphenophyllum Schlotheimii, Sphenopteris Hildreti.

COAL D or E.

Hopkins and Christian Counties.

Calamites cruciatus, C. ramosus, Neuropteris dentata, N. fimbriata, N. flexuosa, N. Loschii, N. tenuifolia, Pecopteris arborescens, P. arguta, P. cyathea, P. notata, P. oreopteridis, P. abbreviata, P. polymorpha, Pseudopecopteris pusilla, Sigillaria fissa, S. obliqua, S. sculpta, Sphenopteris Dubuissonis, S. Gravenhorstii, S. mediana, S. plicata.

Localities of uncertain horizon.

Calamites gracilis, Pecopteris Miltoni, P. nodosa.

ILLINOIS.

COAL B.

Murphysborough, Jackson County.

Alethopteris Serlii, Annularia longifolia, A. sphenophylloides, Calamites ramosus, Carpolithes cistula, C. persicaria, Cordaites borassifolius, Dictyopteris rubella, Lepidocystis vesicularis, Lepidodendron Worthenii, Lepidostrobus foliaceus, Neuropteris aspera, N. capitata, N. fimbriata, N. flexuosa, N. hirsuta, N. Loschii, N. rarinervis, N. tenuifolia, N. vermicularis, Odontopteris heterophylla, O. Worthenii, Pecopteris oreopteridis, P. villosa, Pseudopecopteris callosa, P. nervosa, Rhabdocarpus Jacksonianus, Sigillaria tessellata, Stigmaria ficoides, S. elliptica, Trigonocarpus juglans.

Mazon Creek.

Alethopteris ambigua, A. falcata, A. lonchitica, A.

Pennsylvanica, A. Serlii, Annularia calamitoides, A. inflata, A. longifolia and var. angustifolia, A. sphenophylloides, Asterophyllites anthracinus, A. equisetiformis, A. foliosus, A. longifolius, A. rigidus, A. sublævis, Calamites approximatus, C. cannæformis, C. Cistii, C. dubius, C. major, C. ramosus C. Suckowii, Callipteridium inflatum, C. Mansfieldi, C. neuropteroides, C. Owenii, C. sinuatum, C. Sullivantii, Cardiocarpus mamillatus, Carpolithes corticosus, Cordaianthus baccifer, C. rugosus, Cordaites diversifolius, C. borassifolius, Dictyopteris obliqua, Equisetites occidentalis, Eremopteris artemisiæfolia, Idiophyllum rotundifolium, Lepidocystis bullatus, Lepidodendron aculeatum, var. caudatum, L. Andrewsii, L. clypeatum, L. dichotomum, L. modulatum, L. rigens, L. rimosum, L. Sternbergii, L. vestitum, Lepidophyllum elegans, L. lanceolatum, L. majus, L. rostellatum, L. striatum, Lepidophloios protuberans, Lepidostrobus Aldrichi, L. connivens, L. foliaceus, L. Goldenbergii, L. hastatus, L. lanceolatus, L. lancifolius, L. oblongifolius, L. ornatus, L. ovalifolius, L. princeps, Lycopodites annulariæfolius, L. cavifolius, L. crassus, Macrostachya infundibuliformis, Megaphytum McLayi, Neuropteris angustifolia, N. anomala, N. capitata, N. Carrii, N. Clarksoni, N. Collinsii, N. coriacea, N. crenulata, N. decipiens, N. Desorii, N. Evenii, N. fasciculata, N. fimbriata, N. Germari, N. hirsuta, N. inflata, N. Loschii, N. flexuosa, N. plicata, N. rarinervis, N. rotundifolia, N. tenuifolia and var. heterophylla, N. verbenæfolia, N. vermicularis, Odontopteris affinis, O. aequalis, O. Bradleyi, O. dilatata, O. monstruosa, O. Schlotheimii, O. sphenopteroides, O. subcuneata, O. Worthenii, Oligocarpia Gutbieri, Pecopteris abbreviata, P. arborescens, P. arguta, P. aspidioides, P. Bucklandi, P. Candolliana, P. Carrii, P. Cistii, P. Clarkii, P. Clintoni, P. acutata, P. elegans, P. emarginata, P. Hallii, P. lanceolata, P. longifolia, P. lyratifolia, P. Miltoni, P. oreopteridis, P. ornata, P. pennæformis, P. pteroides, P. robusta, P. serpillifolia, P. solida, P. squamosa, P. stellata, P. Strongii, P. unita, P. venulosa, P. vestita, P. villosa, Pseudopecopteris anceps, P. latifolia, P. Mazoniana, P. muricata, P. nervosa, P. Newberyi, P.

hymenophylloides, P. Pluckneti, P. Sillimanni, P. subcrenulata, Rachiopteris affinis, R. Selago, Rhabdocarpus clavatus, R. cornutus, R. Jacksonianus, R. mamillatus, R. multistriatus, R. pachytesta, Rhacophyllum Clarkii, R. cornutum, R. corallinum, R. filiciforme, R. flabellatum, R. fucoideum, R. Gutbierianum, R. hamulosum, R. hirsutum, R. lactuca, R. molle, R. spinosum, R. Strongii, R. thalliforme, Sigillaria monostigma, S. tessellata, Sigillarioides radicans, Sorocladus sagittatus, S. Worthenii, S. ophioglossoides, Sphenophyllum emarginatum, S. filiculme, S. longifolium, S. Schlotheimii, Sphenopteris chærophylloides, S. cristata, S. elegans, S. plicata, S. pseudo-Murrayana, S. solida, S. subalata, S. tridactylites, Spirangium appendiculatum, S. multiplicatum, S. Prendelii, Stemmatopteris cyclostigma, S. hirsuta, Stigmarioides Evenii, S. linearis, S. truncatus, S. tuberosus, S. villosus, Trigonocarpus Menzelianus, T. Nœggerathii, T. Starkianus.

Morris.

SHALE OVER COAL.

Alethopteris aquilina, A. lonchitica, A. Pennsylvanica, A. Serlii, Annularia longifolia, A. sphenophylloides, Asterophyllites equisetiformis, A. grandis, A. longifolius, A. rigidus, Calamites approximatus, C. cannæformis, C. cruciatus, C. ramosus, Callipteridium Sullivantii, Caulopteris Cistii, C. obtecta, Cordaites diversifolius, C. gracilis, Dictyopteris cordata, D. neuropteroides, Eremopteris artemisiæfolia, Lepidocystis bullatus, L. vesicularis, Lepidodendron dichotomum, L. modulatum, L. Morrisianum, L. obovatum, L. rimosum, L. Sternbergii, L. Veltheimianum, Lepidophloios crassicaulis, L. ichthyoderma, L. laricinus, L. protuberans, Lepidophyllum auriculatum, L. lanceolatum, L. majus, L. Morrisianum, Lepidostrobus foliaceus, L. incertus, L. princeps, Lycopodites Meekii, L. pendulus, Megaphytum Grand'Euryi, M. McLayi, Neuropteris angustifolia, N. anomala, N. callosa, N. cordata, N. Collinsii, N. decipiens, N. fimbriata, N. flexuosa, N. hirsuta, N. rarinervis, N. tenuifolia, N. vermicularis, Odontopteris Schlotheimii, Oligocarpia Alabamensis, O. Gutbieri, Pachypteris gracillima, Pecopteris arguta, P. aspera, P. cristata, P. dentata, P. erosa, P. lyratifolia, P. oreopteridis, P. pennæformis, P. quadratifolia, P. squamosa, P. Strongii, P. vestita, P. villosa, Pseudopecopteris anceps, P. callosa, P. nervosa, P. Pluckneti, Rhabdocarpus multistriatus, Rhacophyllum adnascens, R. arborescens, R. hirsutum, R. thalliforme, Sigillaria lepidodendrifolia, S. monostigma, S. Schimperi, Sorocladus asteroides, Sphenophyllum cornutum, S. emarginatum, S. Schlotheimi, Sphenopteris chærophylloides, S. furcata, S. gracilis, S. mediana, S. mixta, S. scaberrima, S. spinosa, S. splendens, S. trichomanoides, S. tridactylites, Stigmaria amoena, S. stellaris, Ulodendron ellipticum, U. elongatum, U. majus, U. punctatum.

Colchester.

Alethopteris aquilina, Annularia longifolia, A. sphenophylloides, Asterophyllites equisetiformis, Calamites approximatus, C. ramosus, Callipteridium Sullivantii, Caulopteris obtecta, Cordaites borassifolius, C. diversifolius, Knorria Selloni, Lepidodendron diplotegioides, L. obovatum, L. rimosum, L. Sternbergii, Lepidostrobus princeps, Lycopodites cavifolius, L. uncinatus, Neuropteris hirsuta, Odontopteris tenuinervis, Pecopteris pennæformis, P. squamosa, P. unita, P. villosa, Pseudopecopteris anceps, P. irregularis, Rhabdocarpus multistriatus, Rhacophyllum Gutbierianum, R. thalliforme, Sigillaria Brardii, S. monostigma, Stigmaria ficoides, S. umbonata, Sphenophyllum cornutum, S. emarginatum, S. Schlotheimii, Sphenopteris spinosa, S. splendens, Ulodendron ellipticum, U. majus, U. punctatum.

Neelysville, Morgan County.

Lepidophyllum auriculatum, Neuropteris fasciculata.

Marseilles, La Salle County.

Sigillaria corrugata, S. Massiliensis, S. spinulosa.

Grape Creek.

Alethopteris ambigua, A. Gibsoni, A. Pennsylvanica, Asterophyllites equisetiformis, Callipteridium Oweni, C.

Sullivantii, Dictyopteris obliqua, Lepidocystis vesicularis, Lepidophyllum acuminatum, L. majus, Lepidophloios auriculatus, L. macrolepidotus, Macrostachya communis, Neuropteris rarinervis, Pecopteris acuta, P. unita, P. villosa, Pseudopecopteris acuta, Sphenopteris spinosa, Sphenophyllum Schlotheimii.

COAL D.

Duquoin.

Alethopteris aquilina, Annularia longifolia, Asterophyllites equisetiformis, Calamites approximatus, C. cruciatus, C. ramosus, Cordaites borassifolius, C. diversifolius, Lepidodendron forulatum. L. Tijoui, L. radicans, Lepidophloios auriculatus, L. obcordatus, Lepidophyllum auriculatum, Lepidostrobus princeps, Megaphytum McLayi, Neuropteris hirsuta, N. rarinervis, Pecopteris pennæformis, P. unita, P. villosa, Rhacophyllum inflatum, Sigillaria Brardii, S. obliqua, Sphenophyllum emarginatum, S. Schlotheimii, Sphenopteris tridactylites, Stemmatopteris insignis.

St. John's, Perry County.

Alethopteris ambigua, Cordaites diversifolius, Lepidodendron forulatum, L. Tijoui, Lepidophyllum auriculatum, Megaphytum McLayi, Pseudopecopteris spinulosa, Sphenopteris paupercula.

COAL E.

Stark County.

Halonia tortuosa, Neuropteris Griffithii, Rhabdocarpus mamillatus, Trigonocarpus Adamsii, T. Dawesii, T. grandis, T. perpusillus, T. Schultzianus, T. Starkianus, Volkmannia fertilis, Sigillaria monostigma.

Peoria County.

Caulopteris Giffordi, Cyclostigma Kiltorkense, Dechenia striata, Halonia secreta, H. tortuosa, Trigonocarpus Giffordi.

COAL G to M.

Grayville, White County.

Calamites Cistii, Carpolithes fasciculatus, Neuropteris Loschii, N. rotundifolia, N. tenuifolia, Pecopteris Miltoni, P. polymorpha, Rhabdocarpus clavatus, Syringodendron Brongniarti.

Carmi, White County.

Calamites approximatus, C. Suckowii, Megaphitum protuberans, Sigillaria spinulosa, S. tessellata, Stemmatopteris Worthenii.

Vigo County.

Palæophycus divaricatus, P. gracilis, P. Milleri.

New Harmony.

Asterophycus Coxii, Didymophyllum Owenii, Lycopodites strictus, Pecopteris Miltoni, Rhabdocarpus lævis.

Horizons undetermined.

McDonough County.

Rhabdocarpus amygdalæformis.

Vandalia Shaft.

Oligocarpia Gutbieri.

MISSOURI.

COAL B or C.

*Clinton.**

Alethopteris ambigua, A. lonchitica, A. Serlii, Annularia longifolia, A. longifolia var. angustifolia, A. sphenophylloides, Asterophyllites fasciculatus, A. rigidus, Calamites Suckowii, C. cistii, Callipteridium membranaceum, C. Owenii, C. Sullivantii, Cordaites communis, C. diversifolius, Cordáianthus dichotomus, C. gemmifer, Dictyopteris obliqua, Eremopteris Missouriensis, Lepidodendron Brittsii, L. cyclostigma, L. lanceolatum, L. marginatum, L. scutatum,

* Clinton is referable to B or C. I name its species in C with those of Cannelton Pa., on account of their affinity.

Lepidodendron Sternbergii, Lepidophloios dilatatus, L. sigillarioides, Lepidophyllum minus, Lepidostrobus Goldenbergii, Lepidoxylon anomalum, Megaphytum Goldenbergii, Neuropteris angustifolia, N. cordata, N. dilatata, N. flexuosa, N. hirsuta, N. Loschii, N. Missouriensis, N. rarinervis, Odontopteris subcrenulata, O. heterophylla, O. sphenopteroides, Pecopteris arborescens, P. Clintoni, P. cristata, P. dentata, P. erosa, Pecopteris pennæformis, P. vestita, Pinnularia capillacea, Pseudopecopteris acuta, P. irregularis, P. macilenta, P. nummularia, P. obtusiloba, P Sillimanni, Rhacophyllum adnascens, R. arborescens, R. filiciforme, R. fimbriatum, R. hamulosum, R. hirsutum, R. lactuca, R. membranaceum, R. spinosum, Sigillaria fissa, S. Menardi, S. reniformis, S. sculpta, S. spinulosa, Sphenophyllum erosum, S. filiculme, S. longifolium, S. oblongifolium, S. Schlotheimii, Sorocladus ophioglossoides, Sphenopteris Brittsii, S. chærophylloides, S. Dubuissonis, S. furcata, S. Gravenhorstii, S. mixta, S. spinosa, S. splendens, S. tridactylites.

Vernon County.

Conostychus Broadheadi, C. prolifer.

HORIZONS NOT DETERMINED.

Neuropteris tenuifolia, Sphenopteris cristata var. minor, Taonurus Colletti.

KANSAS

Osage City.

Calamites ramosus, Carpolithes granularis, Lesleya microphylla, Neuropteris angustifolia, Odontopteris Reichiana, O. Schlotheimii, Pecopteris Clintoni, Rhabdocarpus amygdalæformis.

Lawrence.

Annularia sphenophylloides, Calamites ramosus, Lesleya microphylla, Neuropteris Griffithii. Sigillariostrobus

Garnett.

Alethopteris Pennsylvanica, Callipteridium membranaceum, Neuropteris Loschii, Oligocarpia Gutbieri, Pecopteris Clintoni, P. erosa.

Thayer.

Knorria compacta, Sphenopteris tenella.

Ottawa.

Alethopteris Pennsylvanica, Calamites ramosus, Neuropteris angustifolia, Pecopteris aspidioides, P. erosa, Pseudopecopteris muricata.

IOWA.

Ottumwa.

Lepidodendron Veltheimianum, Sigillaria Saullii.

Table of Distribution of the Plants Described in the U. S. Coal Flora.

	Devonian.	Chemung.	Catskill.	Pocono and sub-carboniferous.	Sub-conglomerate.	Intra-conglomerate.	Anthracite. Coal A, B, C.	Anthracite. Coal D, E, F.	Anthracite. Coal G to M.	Anthracite. Coal M and above.	Anthracite. Stage unknown.	Bituminous. Coal A, B.	Bituminous. Coal C.	Bituminous. Coal D, E, F.	Bituminous. Coal G to M.	Bituminous. Stage unknown.	Found in Europe.
	1	*2*	*3*	*4*	*5*	*6*	*7*	*8*	*9*	*10*	*11*	*12*	*13*	*14*	*15*	*16*	*17*
1. CRYPTOGAMOUS,																	
1. FUNGI,																	
3. *Rhizomorpha*,																	
3. R. Sigillariæ,													1				
4. THALLASSOPHYTES,																	
699. *Dendrophycus*,																	
699. D. Desorii,					3												
6. *Taonurus*,																	
7. T. marginatus,						1											
7. T. Colletti,					1												
8. T. Cauda Galli,	1	1															
9. *Palæophycus*,																	
10. P. Milleri,															1		
11. P. gracilis,															1		
11. P. divaricatus,															1		
12. *Asterophycus*,																	
12. A. Coxii,															1		
13. A. simplex,															1		
14. *Conostychus*,																	
15. C. Broadheadi,					1												
16. C. prolifer,						1											
17. C. ornatus,					2	1											
Dictyophytum,																	
D. ramosum,			1														
Acanthophytum,																	
A. spinosum,		1															
17. CALAMARIEÆ,																	
19. *Calamites*,																	
20. C. Suckowii,					2		2	3	1	1		5	2		1		+
21. C. major,							1	1		1		1					+
22. C. ramosus,					2	1	1	3				3	1	2		2	+
23. C. ramifer,					1	1							1			1	+
24. C. cannæformis,					2		2	2	1	1		2	1	1			+
25. C. gigas,													1				+
26. C. approximatus,					2			5				3	1	1	1		+
26. C. cruciatus,					2			1	1			1					+
27. C. Cistii,					1		1	3				2	1	1	1		+
27. C. dubius,								3	1			1				1	+
28. C. pachyderma,						1											+
29. C. disjunctus,										1			1		1		
29. C. gracilis,																1	
C. inornatus,	1																
30. *Bornia*,																	
30. B. radiata,	1				4	1										1	

		1	2	3	4	5	6	7	8	9	10	11	12	13	14	15	16	17
32.	*Calamodendron?*																	
708.	C. species,													1				
34.	*Asterophyllites*,																	
35.	A. equisetiformis,					3	1	2	2	1	1	1	4	1	1			+
	A. " var. erectifolius,					1												
36.	A. anthracinus,								1				1	1				+
36.	A. longifolius,						1		1				2					+
37.	A. rigidus,						1		1			1	2	2				+
38.	A. sublævis,											1	1	1				
38.	A. foliosus,					1	1						1	1				+
41.	A. grandis,					1						1	1	1				+
41.	A. fasciculatus,													1				
714.	A. gracilis,					4												
43.	A. minutus,					1												
44.	*Annularia*,																	
45.	A. longifolia,						1	2	4			1	6	2	1	1	1	+
	A. " " var. augustifolia,							1	2	1			1	1				
47.	A. inflata,							1	3			2	2	1				
48.	A. calamitoides,										1	1	1					+
48.	A. sphenophylloides,						1	1	1			2	4	2			2	+
724.	A. " var. intermedia,																1	
725.	A. minuta,										1							+
50.	A. radiata,					1			1									+
725.	A. Emersoni,															1		
51.	A. Dawsoni,					1										1		
725.	A. cuspidata,					1												
723.	A. tuberculata,						1						2					+
51.	*Sphenophyllum*,																	
52.	S. Schlotheimii,					2	1	2	7			1	5	2	2	1	1	+
53.	S. emarginatum,						1		1		1	1	4	1	3			+
53.	S. longifolium,					1			2			1	2	2		1		+
	S. " var. latifolium,					1												
55.	S. erosum,													1				+
55.	S. bifurcatum,					2												
56.	S. cornutum,								2				2					
57.	S. oblongifolium,													2				+
58.	S. filiculine,										2	1	1	1				
726.	S. augustifolium,														1	1	1	+
726.	S. saxifragæfolium,					1												+
728.	S. tenerrimum,					1												+
	S. antiquum,	1																
50.	*Calamostachys*,																	
715.	C. lanceolata,					2												
717.	C. ovalis,					1							3					
718.	C. brevifolia,													1				
719.	*Volkmannia*,																	
720.	V. prælonga,							1										
720.	V. fertilis,							1							1			
719.	V. crassa,					1												
60.	*Macrostachya*,																	
60.	M. infundibuliformis,					1		1				1		1				+
828.	M. communis,									1			1					
721.	M. lanceolata,						1											
829.	M. minor,						1											
829.	M. aperta,										1			1				
62.	*Equisetites*																	
62.	E. occidentalis,												1	1				
729.	E. gracilis,					1												
65.	FILICACEÆ,																	
73.	NEUROPTERIDEÆ,																	
75.	*Neuropteris*,																	
77.	*N. Cylopterides*,																	
77.	N. reniformis,					1			1				2	1	1			+
78.	N. dilatata,													1				+

		1	2	3	4	5	6	7	8	9	10	11	12	13	14	15	16	17
79.	N. trichomanoides,								1		1			1				+
80.	N. laciniata,											1						
	Nephropterides,																	
81.	N. fimbriata,					2			4	1	2		3	1	3	1		
82.	N. dentata,										2	1			2			
83.	N. Rogersi,										1			1				
84.	N. gibbosa,										1			1				
85.	N. auriculata,													1				+
86.	N. inflata,					1		1	2	1			1	1				
87.	N. Collinsii,								1	1			2	1				
731.	N. Carrii,								3				1					
	Euneuropterides,																	
88.	N. hirsuta,					2		1	2	1	1		8	2	1	1		+
89.	N. augustifolia,							2	4		1		2	3	1		1	+
	var. hirsuta,																1	
737.	N. Blissii,												1					+
91.	N. cordata,								2		1		2	2			1	+
733.	N. decipiens,												1	1				
93.	N. fasciculata,												2	1				
94.	N. Clarksoni,								2	1	1		3	1				
96.	N. plicata,								6		1		1	1	1	1		+
97.	N. rotundifolia?							1	2		1		1	1		1		+
98.	N. Loschii,					1			3	1			3	2	2	2	1	+
96.	N. flexuosa,												3	1	3			+
99.	N. vermicularis,								3	1			5	1				
100.	N. tenuifolia,					4			1		1		2	2	2	2	1	+
	var. heterophylla,								2									
732.	N. oblongifolia,						1											
102.	N. subfalcata,					1												
103.	N. capitata,					1			1				2	1				
104.	N. Missouriensis,													1				
105.	N. Grangeri,								1		3			1			2	+
	Pachydermateæ,																	
734.	N. Smithii,					5												
735.	N. Elrodi,					4									1			
108.	N. obscura,											2			1	1		
109.	N. rarinervis,								3	1		1	6	2	1			+
111.	N. coriacea,							1	1				1	1				
112.	N. Desorii,							1	1	1	2		1	1				
113.	N. Germari,					2			1		1		1	1				+
737.	N. Griffithii,								1						1		1	
115.	N. callosa,								3		1	1	1			1		
116.	N. crenulata?								2	1	1	1	1	1		1		+
117.	N. Evenii,												1					
117.	N. Agassizi,											1		1				
	Anomales,																	
118.	N. anomala,										1		2	1		1		
120.	N. verbenæfolia,												1	1				
121.	N. biformis,					5												
736.	N. aspera,						1						1	1				
122.	N. fissa,										1							
123.	N. minor,											1						
123.	N. acuminata,											1						
837.	Cyclopteris Brownii,		1															
830.	" incerta,	1																
829.	" valida,			1														
124.	*Odontopteris*,																	
125.	O. tenuinervis,										1		1					
126.	O. Alpina,								1	1		1						+
127.	O. Newberryi,					2	2											
128.	O. cornuta,													1				
129.	O. heterophylla,												2	1				
130.	O. Worthenii,									1			2					
131.	O. alata,										2							

	1	2	3	4	5	6	7	8	9	10	11	12	13	14	15	16	17
132. O. Brardii,					1						1		1				
830. O. Britannica,													1		1		+
133. O. squamosa,											1	1					
134. O. subcuneata,												2	1	1			
831. O. dilatata,												1					
743. O. æqualis,												1	1				
136. O. Schlotheimii,							1	2		2	1	4	1	1	1	1	+
137. O. subcrenulata,										2	1		1				
138. O. abbreviata,							1						2				
139. O. sphenopteroides,					1							1	1				
140. O. gracillima,						1											
140. O. Brardleyi,												1			1		
743. O. deformata,											1						
740. O. neuropteroides,					1						1						
740. O. patens,											1		1				
831. O. Reichiana,							1									1	+
742. O. affinis,												1	1				
741. O. monstruosa,												1	1				
142. *Lesleya*,																	
143. L. grandis,					1												
831. L. microphylla,																2	
143. *Dictyopteris*,																	
145. D. rubella,					1							1					
146. D. obliqua,					1		2	4		1	1	4	2			1	
832. D. Scheuchzeri,								1					1				+
833. D. neuropteroides,												1					+
833. D. cordata,												1					+
147. *Megalopteris*,																	
148. M. Southwellii,					1												
149. M. Hartii,					1												
149. M. minima,					1												
M. Sewelleni,					1												
149. M. ovata,					1												
150. M. fasciculata,					2												
151. M. abbreviata,					1												
151. M. lata,					1												
152. M. marginata,					1												
744. M. rectinervis,					1												
834. M. serrata,					1												
833. M. dentata,					1												
151. M. Dawsoni,				1													
153. *Tæniopteris*																	
153. T. Smithii,					1												
743. T. truncata,						1											
154. *Neriopteris.*																	
154. N. lanceolata,						2											
155. *Orthogoniopteris.*																	
156. O. clara.,					1												
165. O. Gilberti,					1												
Danæites.																	
157. D. Emersoni.															1		
158. D. macrophyllus,					1	1					1						
159. *Idiophyllum.*																	
160. I. rotundifolium,												1					
Alethopterides.																	
162. *Lescuropteris.*																	
162. L. Moorii,										1					2		
163. L. adiantites,																	
164. *Callipteridium.*																	
164. C. Sullivantii,					1		1	1	1			6	2				
166. C. Mansfieldi,												1	1				
166. C. neuropteroides,													1				
167. C. Owenii,					1			1				5	1				
745. C. inæquale,												1	1				

	1	2	3	4	5	6	7	8	9	10	11	12	13	14	15	16	17
169. C. Pardeei,							?										
169. C. rugosum,					1			1	1	1							
171. C. Aldrichi,					2												
172. C. membranaceum,													1				
173. C. Massilioneum,								1				1					
174. C. inflatum,												1					
746. C. rigidum,								1									
747. C. Dournaisii,						1											+
748. C. Grandini,								1					1				+
745. C. sinuatum,									1			3					+
175. *Alethopteris.*																	
176. A. Serlii,					2		2	8			1	10	2			1	+
177. A. lonchitica,					1	1		3			2	3	2				+
T. " var. Davreuxii,						1											+
179. A. grandifolia,					1	1											
179. A. Helenæ,				1	5	1											
181. A. Pennsylvanica,								2				3	1			2	+
181. A. aquilina,											1	3		1			+
182. A. ambigua,											1	3	2	1			
183. A. Gibsoni,												1			1		
185. A. Bunburyi,					1												
186. A. falcata,												1					
186. A. marginata,										1	1						+
834. A. Evansii,					1												
750. A. maxima,					1												
748. A. crassa,													1				
835. A. robusta,													1				
188. *Protoblechnum.*																	
188. P. Holdeni,					1												
PSEUDOPECOPTERIDEÆ.																	
190. *Pseudopecopteris.*																	
190. P. Mazoniana,												1					
193. P. subcrenulata,												1	1				
194. P. Sheaferi,										1							
195. P. spinulosa,											1			1			
196. P. hymenophylloides,												1					
197. P. nervosa,					3	1	2	4	1			5	1				
P. " var. Sauveurii,													1				+
198. P. sub-nervosa,													1				+
199. P. Pluckneti,					1			4	1		1	3	1	1			+
755. P. macilenta,					2								2				+
755. P. hispida,													1				
751. P. dimorpha,								1	1		1						
202. P. Newberryi,							1	1	1			3					
203. P. abbreviata,										1							
203. P. muricata,					5	1				1	1	1				1	+
205. P. cordato-ovata,					1			2	1								+
206. P. Sillimanni,						1						3		1			+
206. P. Loschii,						1				1							+
207. P. anceps,					1			4			2	5	1	1			
751. P. decurrens,								1		1							
209. P. callosa,												2					
210. P. glandulosa,						1					1						
211. P. irregularis,					1	1		1				1	1				+
752. P. nummularia,					1	1		1					2		1		+
212. P. denudata,								1					1				
214. P. decipiens,					3			1			1			1			
215. P. latifolia,					5	1		1	1	1		5	1				+
215. P. acuta,												1	1				+
216. P. speciosa,					2												
217. P. Virginiana,				1													+
217. P. trifoliata,					1												
218. P. polyphilla,					1												+
220. P. pusilla,										1				2			+

	1	2	3	4	5	6	7	8	9	10	11	12	13	14	15	16	17
264. P. incompleta,	.	.	.	.	.	.	.	.	.	.	.	.	1	.	.	.	.
753. P. obtusiloba,	.	.	.	1	1	.	.	.	.	.	.	.	1	.	.	.	+
754. P. Andræana,'	.	.	.	.	.	.	.	.	.	.	.	.	.	.	1	.	.
PICOPTERIDEÆ.																	
223. *Pecopteris.*																	
223. (*Goniopteris.*)																	
223. P. unita,	.	.	.	.	.	.	.	4	1	1	2	5	1	2	.	.	+
225. P. emarginata,	.	.	.	.	.	.	.	1	.	.	.	1	.	.	.	.	+
226. P. longifolia,	.	.	.	.	.	1	.	2	.	.	.	1	.	.	.	.	+
227. P. lanceolata,	.	.	.	.	.	.	.	1	.	.	.	1	.	2	.	.	.
227. P. arguta,	.	.	.	.	.	.	.	2	1	1	1	2	.	.	.	.	+
228. P. elegans,	.	.	.	.	.	.	.	.	1	1	.	.	.	.	.	.	+
756. P. robusta,	.	.	.	.	.	.	.	.	1	1	.	.	.	.	.	.	+
230. P. venulosa,	.	.	.	.	.	.	.	.	.	.	.	2	.	.	.	.	.
(*Cyatheites.*)																	
230. P. arborescens,	.	.	.	.	.	.	.	1	1	3	1	2	.	2	3	.	+
231. P. cyathea,	.	.	.	.	.	.	.	.	.	1	.	.	.	2	.	.	+
232. P. platyrachis,	.	.	.	.	.	.	.	.	.	1	1	.	.	.	.	.	+
756. P. aspidioides,	.	.	.	.	.	.	.	1	.	.	1	1	.	.	.	1	+
233. P. nodosa,	.	.	.	.	.	.	.	.	.	1	.	.	.	.	2	1	+
234. P. quadratifolia,	.	.	.	.	.	.	.	.	1	.	1	1	1	.	.	.	.
235. P. squamosa,	.	.	.	.	.	.	.	.	.	.	.	4	1	.	1	.	.
236. P. Strongii,	.	.	.	.	.	.	.	.	.	.	.	2	1	1	.	.	.
237. P. serpillifolia,	.	.	.	.	.	.	.	.	.	.	.	1	.	.	.	.	.
238. P. oreopteridis,	.	.	.	.	.	.	.	4	.	1	2	4	2	3	2	.	+
239. P. pennæformis,	.	.	.	.	.	.	3	4	1	1	1	3	2	2	.	.	+
240. P. dentata,	.	.	.	.	.	1	.	5	1	.	1	1	2	.	.	.	+
241. P. acuta,	.	.	.	.	1	.	.	1	1	.	.	1	2	.	.	.	+
757. P. serrulata,	.	.	.	.	2	.	.	.	.	.	.	.	.	.	.	.	.
242. P. aspera,	.	.	.	.	.	.	.	.	.	.	.	2	.	.	.	.	+
243. P. Candolliana,	.	.	.	.	.	.	.	4	.	.	1	1	.	.	.	.	+
243. P. Cistii,	.	.	.	.	.	.	.	3	1	.	.	1	1	.	.	.	+
244. P. Bucklandi,	.	.	.	.	.	.	.	2	.	.	.	1	1	.	.	.	+
245. P. elliptica,	.	.	.	.	.	1	.	3	1	.	1	.	.	1	.	.	.
246. P. distans,	.	.	.	.	.	.	.	.	.	1	1	.	.	.	.	.	.
247. P. Miltoni,	.	.	.	.	.	.	.	3	.	2	2	.	1	1	4	1	+
247. P. polymorpha,	.	.	.	.	.	.	.	.	1	1	.	.	1	2	1	.	+
248. P. abbreviata,	.	.	.	.	.	.	.	2	.	1	.	1	.	2	.	.	+
835. P. Schimperi,	.	.	.	.	.	.	.	.	.	.	.	.	.	1	.	.	.
249. P. pteroides,	.	.	.	.	1	.	.	1	1	.	.	1	.	.	1	.	+
758. P. obsoleta,	.	.	.	.	.	.	.	.	.	.	.	.	1	.	.	.	.
(*Villous.*)																	
250. P. velutina,	.	.	.	.	.	.	.	.	.	.	1	.	1				
251. P. Clintoni,	.	.	.	.	.	.	.	1	.	.	.	1	3	.	.	2	.
252. P. vestita,	.	.	.	.	.	.	1	1	1	.	.	2	2	.	.	.	.
253. P. villosa,	.	.	.	.	.	.	.	4	1	.	.	6	1	1	.	.	+
(*Crestate.*)																	
255. P. erosa,	.	.	.	.	1	.	.	1	.	.	1	1	2	.	.	3	+
256. P. cristata,	.	.	.	.	.	.	.	.	.	.	.	1	2	.	.	.	+
759. P. Georgiana,	.	.	.	.	2	.	.	.	.	.	.	.	.	.	.	.	.
256. P. serrula,	.	.	.	.	.	.	.	.	.	.	1	.	.	.	.	.	+
257. P. angustissima?,	.	.	.	.	1	.	.	.	.	.	.	.	.	.	.	.	.
258. P. Hallii,	.	.	.	.	1	.	.	.	.	.	.	1	.	.	.	.	.
259. P. lyratifolia,	.	.	.	.	.	.	.	.	.	.	.	2	1	.	.	.	+
260. P. stellata,	.	.	.	.	.	.	.	.	.	.	.	1	.	.	.	.	.
760. P. ornata,	.	.	.	.	.	.	.	.	.	.	.	1	.	.	.	.	.
261. P. solida,	.	.	.	.	.	.	.	.	.	.	.	1	.	.	.	.	.
261. P. Clarkii,	.	.	.	.	.	.	.	.	.	.	1	1	.	.	.	.	.
262. P. notata,	.	.	.	.	.	1	1	.	.	1	.	.	.	2	.	.	.
759. P. Carrii,	.	.	.	.	.	.	.	.	.	.	.	1	.	.	.	.	+
263. P. mycrophylla,	.	.	.	.	.	.	.	.	.	.	.	1	.	.	.	.	.
264. P. concinna,	.	.	.	.	.	.	.	.	.	1	.	.	.	.	.	.	.
265. *Oligocarpia.*																	
266. O. Alabamensis,	.	.	.	.	1	.	.	.	.	.	.	1	.	.	.	.	.

	1	2	3	4	5	6	7	8	9	10	11	12	13	14	15	16	17
266. O. Gutbieri,								1				2				2	+
267. O. flagellaris,										1							
268. SPHENOPTERIDEÆ.																	
269. (*Pecopterids.*)																	
269. S. goniopteroides,													2				
270. S. chærophylloides,					1			1			1	2	2				+
271. S mediana,												3		2			
271. S. pseudo-Murrayana,					3						1	1					
762. S. communis,					3												
272. S. subalata,												1	1				+
761. S. cristata,					1						1	1	1				+
762. S. " var. minor,					1												+
835. S. crenata.					1												+
S. (*Proper.*)																	
763. S. plicata,											1	1		2			
763. S. Gravenhorstii,					1			1			1		1	2			+
763. S. " var. B.,					1												
275. S. Dubuissonis,													1	2	1		+
764. S. hymenophylloides,													1				
220. S. adiantoides,					1												+
276. S. mixta,					3							1	2	1			
276. S. gracilis,												1					
765. S. Birttsii,													1	1			
278. S. paupercula,										1				1			
279. S. scaberrima,												1					
280. S. microcarpa,					1												
765. S. inæquilateralis,					1												
(*Hymenophyllites.*)																	
281. S. spinosa,					1							3	1	1			+
769. S. solida,												1					
282. S. splendens,												2	1				
282. S. furcata,					2	1		1				3		1			+
283. S. Hildreti,												4					
768. S. Royi,												1					
284. S. flexicaulis,					2												
284. S. tridactylites,					4			1			1	4	2				+
766. S. Harveyi,					1												
286. S. trichomanoides,					1		1					1					+
286. S. quercifolia,					1												+
287. S. elegans,					2						1	1					+
767. S. divaricata,					1									1			+
288. S. Larischii,					2												+
767. S. Tracyana,					1												
288. S. Hoeninghausii,					6						1						+
770. S. obovata,					1												
290. S. Ballantini,					1												
290. S. linearis,						1	1										+
291. S. flaccida,				1													+
836. S. tenella,																1	+
836. S. dissecta,																1	
837. S. fascicularis,									1								
837. S. Hitchcockiana,			1														
296. S. marginata,					1												
292. *Eremopteris.*																	
292. E. crenulata,					2												
293. E. flexuosa,					2												
293. E. dissecta.					1												
293. E. artemisiæfolia,					2				1			4	1				+
294. E. elegans,					2								1				+
770. E. Cheathami,					1												
295. E. Missouriensis,													1				
296. E. (triphyllopteris) microphylla,					2												

	1	2	3	4	5	6	7	8	9	10	11	12	13	14	15	16	17
ADIANTITES.																	
297. *Triphyllopteris.*																	
297. T. Lesceuriana,				1													
298. *Archæopteris.*																	
300. A. obliqua,				1													
301. A. obtusa,			2	1													
302. A. minor,			2	1													
304. A. Halliana,		1		1	1												
776. A. Hybernica,			2														
306. A. Bockschiana,				3													
304. A. Jacksoni,		1	2														
776. A. Rogersi,			1														
774. A. denticulata,					1												
303 A. stricta,					1												
307. A. Alleghaniensis,					1												
775. A. sphenophyllifolia,			1														
775. A. macilenta,			1														
307. *Crematopteris.*																	
307. C. Pennsylvanica,															1		
308. *Pachypteris.*																	
308. P. gracillima,												1	1				
Rhacophyllum.																	
310. (*Neuropterides.*)																	
311. R. flabellatum,					2												+
311. R. truncatum,			1	1													
312. R. membranaceum,													1				
313. R. scolopendrites,															1		
(*Hymenophyllites.*)																	
314. R. arborescens,												1	1				
315. R. lactuca,					1			2		1		1	2				+
316. R. filiciforme,					1		1	3	1		2		2	1			+
838. R. filiforme,											1					1	+
317. R. corallinum,											1						
317. R. cornutum,												1	1				
318. R. hirsutum,										1	1	1	1				
318. R. fimbriatum,										1	1		1				
R. " var. affine,										1			1				
319. R. Clarkii,									1			1	1				
320. R. Gutbierianum,						1		2				3	1				+
320. R. spinosum,												1	1				
321. R. hamulosum,												1	1				
321. R. adnascens,					1							1	2				+
322. R. trichoideum,								1									
323. R. inflatum,														1			
323. R. expansum,									1				1				
324. R. thalliforme,												4					
Fucoides.																	
325. R. fucoideum,												1					
325. R. Strongii,												1					
326. R. molle,												1					
326. R. irregulare,																1	+
327. *Sorocladus.*																	
328. S. stellatus,					1												
328. S. asteroides,												1					
329. S. sagittatus,												1					
329. S. ophioglossoides,												1	1				
330. S. Worthenii,												1					
331. *Rachiopteris.*																	
331. R. affinis,												1					
332. R. selago,												1					
332. R. pinnata,	1																
332. R. cyclopteroides,			1														
332. R. punctata,			1														
333. R. striata,	1	1															

		1	2	3	4	5	6	7	8	9	10	11	12	13	14	15	16	17
333.	R. tenuistriata,		1															
	R. squamosa,													1				
333.	*Stigmarioides.*																	
333.	S. Evenii,												1					
334.	S. truncatus,												1					
334.	S. villosus,												1					
335.	S. tuberosus,												1					
335.	S. linearis,												1					
337.	*Stemmatopteris.*																	
337.	S. hirsuta,									1			1	1				
337.	S. emarginata,													1				
338.	S. Schimperi,													1				
339.	S. squamosa,									1				1				
339.	S. angustata,													1				
339.	S. punctata,								1	1	1			1				
339.	S. peltigera,									1								+
340.	S. insignis,													1	1			
340.	S. gigantea,							1						1				
341.	S. cyclostigma,									1			1	1				
341.	S. mimica,													1				
342.	S. polita,													1				
342.	S. Worthenii,															1		
838.	S. microstigma,													1				
838.	S. anceps,													1				
	Caulopteris.																	
343.	C. Giffordi,														1			
344.	C. Lacoei,								1	1								
344.	C. obtecta,								3				2	1				
345.	C. Cistii,							1	1				1	1				+
346.	C. Mansfieldi,								1	1				1				
347.	C. Lockwoodi,		1															
347.	C. antiqua,		1															
348.	C. peregrina,		1															
	Megaphytum.																	
349.	M. McLayi,												2	1	2			
349.	M. Goldenbergii,													1				+
350.	M. Grand'Euryi,									1			1					
352.	M. protuberans,															1		
	Psaronius.																	
353.	P. species,															1		
	LYCOPODIACEÆ.																	
	Lycopodites et Selaginites.																	
778.	L. arborescens,					1		1										
357.	L. pendulus,					1							1					
357.	L. Meekii,					1							1					
358.	L. cavifolius,												4					
358.	L. crassus?,												1					
779.	L. flexifolius,					1												
359.	L. uncinatus,												1					
359.	L. Ortoni,														1			
360.	L. strictus,															1		
780.	L. Lacoei,								1	1								
779.	L. simplex,					1	1											
361.	L. annulariæfolius,												1					
362.	L. Richardsoni,		1															
362.	L. comosus,		1															
363.	*Lepidodendron.*																	
366.	L. Sternbergii,					4	2	2					6	2				+
368.	L. Brittsii,								1					1				
369.	L. lanceolatum,													1				
369.	L. scutatum,							1						1				
370.	L. latifolium,								1	1								
370.	L. Morrisianum,												1					
371.	L. aculeatum,					2		4	4	1		1	1	1				+

	1	2	3	4	5	6	7	8	9	10	11	12	13	14	15	16	17
L. aculeatum Var. caudatum,													1				+
372. L. rigens,												1	1	1			
373. L. longifolium,					2			1									+
374. L. Veltheimirnum,					7	3	3	5	1			1				1	+
376. L. squamiferum,					1												
377. L. primæve,	1																
377. L. corrugatum,	1	1		1	3												
379. L. vestitum,					2		3	1				1	1				
377. L. Rushvillense,					5												
380. L. clypeatum,					3	1	4	2	1			1					
381. L. costatum,					1												
382. L. turbinatum,					1												
382. L. rhombicum,					1			1									+
383. L. quadrangulatum,																1	
382. L. quadratum,													1				+
384. L. dichotomum,					2	1	4	1			1	3	1				+
var. obovatum,					2	1	3	3				2			1		+
385. L. modulatum,					1	2	3	3	1			2	1				
386. L. carinatum,							1								1		
387. L. distans,							1										
388. L. cuspidatum,								1								1	
388. L. Worthenii,												1					
389. L. Andrewsii,												1					
389. L. quadrilaterale,					1												
390. L. forulatum,														2			
390. L. diplotegioides,					1	1						1					
391. L. Tyoni,							2							2			
392. L. obtusum,							1	1									
392. L. rimosum,						1	3	2			1	5	1				+
394. L. crenatum,					1												+
394. L. cyclostigma,													1				
395. L. Mielicki,								1									+
395. L. Gaspianum,			2		1												
396. L. Chemungense,		1															
396. L. ichthyolepis,											1						
397. L. obscurum,					1												
397. L. radicans,														1			
382. L. Volkmannianum,					2			1									+
784. L. marginatum,															1		+
839. L. rigidum,					2												
397. *Ulodendron.*																	
401. U. commutatum,					1												+
401. U. majus,					1	1		2				2		1		1	+
403. U. minus,					3	1		2									+
404. U. ellipticum,							1					2					+
405. U. elongatum,												1					
405. U. punctatum,							1	2				2	1				
407. *Knorria.*																	
407. K. imbricata,					2		1	2			1						+
839. K. compacta,																1	
839. K. Selloni,												1					+
408. K. acicularis,				1			1										+
409. *Halonia.*																	
411. H. tuberculata,					1	1											+
413. H. tortuosa,								1						2			+
414. H. Mansfieldi,													1				
416. H. flexuosa,																1	+
417. H. pulchella,					1												
417. H. secreta,														1			
418. *Lepidophloios.*																	
420. L. crassicaulis,					2		1						1				+
421. L. auriculatus,								1						1			
422. L. laricinus,					3	1		1	1			1	1				+
423. L. obcordatus,														1			

	1	2	3	4	5	6	7	8	9	10	11	12	13	14	15	16	17
424. L. macrolepidotus,												1					+
425. L. sigillarioides,													1				
425. L. protuberans,												2					
426. L. ichthyoderma,												1					
781. L. dilatatus,													2				
427. *Fructifications of Lepidophilois,*																	
428. L. species,												2					
429. *Cyclostigma.*																	
429. C. Kiltorkense,														1			+
430. *Dechenia.*																	
431. D. striata,														1			
431. *Lepidostrobus.*																	
432. L. Goldenbergii,					1				1			1	2				+
433. L. praelongus,							1		1								
434. L. princeps,												5		1			
434. L. variabilis,						1		1	1				1				+
840. L. Butleri,								1									
435. L. spectabilis,													1				
436. L. lancifolius,								1				1					
436. L. lanceolatus,						1	1	5	1		2	2					
841. L. latus,									1								
437. L. oblongifolius,							1	1			1	1					
438. L. ovatifolius,							2	2				2					
438. L. hastatus,						1					1	1					
440. L. ornatus,							1	1	1			1	1				+
441. L. Aldrichi,					1							1					
441. L. connivens,						1						1					
442. L. species,												1					
443. L. incertus,												1					
443. *L. Macrocystis.*																	
443. L. (M.) Salisburyi,												1					
444. L. (M.) quadratus,													1				
444. L. (M.) Mansfieldi,													1				
445. L. (M.) foliaceus,												2	1				
446. L. (M.) mirabilis,						1											
784. L. truncatus,													1				
L. Richardsoni,			1														
L. globosus,			1														
447. *Lepidophyllum.*																	
447. L. affine,							1	3		1							
447. L. brevifolium,					1		2					1					
788. L. tumidum,								1			1						
448. L. Morrisianum,												1					
449. L. majus,								1				3	1				+
449. L. Mansfieldi,								1	1				1				
450. L. auriculatum,												2	2				
450. L. acuminatum,					1	1	2	1	1			1	1	1			
451. L. obtusum,							1										
451. L. rostellatum,								1	1			1					
452. L. striatum,								4				1					
452. L. linearifolium,							2										
787. L. coriaceum,													1				
785. L. cultriforme,						2					1		1				
786. L. fallax,													1				
436. L. lanceolatum,							2	2									
787. L. elegans,								1					1				
786. L. Campbellianum,					1	1	1										
787. L. minutum,							1										
786. L. gracile,					1												
841. L. Stantoni,								1			1						
454. *Lepidocystis.*																	
454. L. pectinatus,						1											
454. L. lineatus,										1							

	1	2	3	4	5	6	7	8	9	10	11	12	13	14	15	16	17
455. L. quadrangularis,							1	2									
455. L. obtusus,					1	1		1	1								
456. L. angularis,						1		2					1				
457. L. vescicularis,						1						3	1				
457. L. fraxiniformis,			1		1	1							1				
458. L. bullatus,					1	1						2					
458. *Sporocystis.*																	
458. S. planus,						1											
Psilophyton.																	
459. P. princeps,	1		2														
460. *Leptophlœum.*																	
460. L. rhombicum,	1																
461. *Tæniophyllum.*																	
464. T. decurrens,													1				
465. T. contextum,													1				
465. T. deflexum,													1				
788. T. brevifolium,													1				
789. *Ptilophitum.*																	
790. P. Vanuxemi,		1			1												
790. P. gracile					1												
791. P. lineare,					2												
466. SIGILLARIEÆ.																	
468. *Sigillaria (Leiodermariæ.)*																	
793. S. monostigma,					1		2					2	2	1			
795. S. Grand'Euryi,													1				
470. S. fissa,											1		1	2			
470. S. sculpta,													1	2			
470. S. obliqua,								1		1				3			+
471. S. spinulosa,												2	1	1	1		+
472. S. dilatata,							1				1						
473. S. reticulata,					2												
473. S. Lorenzii,								1									
474. S. stellata,							1					2					
474. S. Schimperi,											1					1	
475. S. corrugata,												3					
476. S. leioderma,											1	1					+
477. S. lepidodendrifolia,							1					1					+
477. *S. Clathrariæ.*																	
797. S. Brardii,					1	1	1	5			2	1	1	2	1		+
479. S. Menardi,						2	1	5			1	1	1		1		+
480. S. Serlii,								2									+
842. S. Brochanti,								1									+
480. *S. Rhytidolepis.*																	
480. S. Dournaisii,					1												
481. S. tessellata,						1	2	3				5	1	1	1		+
482. S. ichthyolepis,														1		1	+
483. S. hexagona,					1											1	+
799. S. mamillaris,					1			3					1				+
485. S. Lescurii,					2		1	2			1						
486. S. notata,										1							+
486. S. cuspidata,								1									+
487. S. Massilliensis,												1					
488. S. attenuata,						1					1	2					
801. S. Williamsii,									1								
489. S. leptoderma,'								1									
490. S. pulchra,						2											
490. S. polita,							1										
491. S. Yardleyi,							1	2			1		1				
491. S. orbicularis,								2					1				+
492. S. Volzii,								1									+
493. S. Pittstoniana,								1									
493. S. Sillimanni,								2									+
494. S. elliptica,						1		2					1				+
495. S. ovalis,								1									

	1	2	3	4	5	6	7	8	9	10	11	12	13	14	15	16	17
495. S. Cortei.							1										+
496. S. obovata,							1										
496. S. acuminata,						1											
497. S. rugosa,								1									+
498. S. marginata,								1									
499. S. Lacoei,								1									
499. S. descoidea,							1					2					
500. S. lævigata,								3									+
501. S. reniformis,								2		1	1	2	2				+
799. S. pyriformis,								1									+
800. S. Leveretti,																1	
842. S. Saullii ?,								1								1	+
842. S. venosa,								2									+
505. S. simplicitas,	1																
505. S. Vanuxemi,	1																
502. *Syringodendron.*																	
502. S. Porteri,																1	
503. S. pachyderma,						1	1	2				1	1				+
504. S. Brongniarti,							1						1		1		+
505. S. cyclostigma,							1	1					1				+
506. S. gracile,		1															
506. *Didymophyllum.*																	
506. D. reniforme,	1																
801. D. Owenii,															1		
507. *Stigmaria.*																	
514. S. ficoides,					1	1	1					2	1				+
516. S. umbonata,								1				1					
516. S. amœna,								1				1					
516. S. stellaris,												1					
377. S. minuta,				2													
843. S. elliptica,											1						
843. S. exigua,		1															
797. *Sigillariostrobus* species.																	
517. *Sigillarioides.*																	
517. S. radicans,								1				1					
518. *Pinnularia.*																	
518. P. capillacea,					1							1	3				
518. *Spirangium.*																	
519. S. Prendelii,												2					
520. S. appendiculatum,						1						1					
520. S. multiplicatum,						1						1					
521. S. intermedium,						1											
521. *Noeggerathieæ.*																	
523. *Whittleseya.*																	
523. W. elegans,						1											
524. W. integrifolia,					1												
525. W. undulata,					1												
843. W. microphylla,					1												
525. *Cordaiteæ.*																	
527. *Cordaites.*																	
529. C. validus,							1	2					1				
530. C. crassus,								1	1								
530. C. grandifolius,						1											
532. C. borassifolius,						1	1	1				5	1				+
533. C. lingulatus,							1										
534. C. communis,					1								1				
535. C. Lacoei,								1									
535. C. diversifolius,					1			2		1		3	1	2			
537. C. Mansfieldi,													1				
539. C. gracilis,												1	1				
540. C. radiatus,													1				
540. C. costatus,													1				
541. C. serpens,													1				
543. C. Robbii,	1				1												

		1	2	3	4	5	6	7	8	9	10	11	12	13	14	15	16	17
544.	C. angustifolius,	1																
544.	C. flexuosus,	1																
545.	*Cordaianthus.*																	
545.	C. ovatus,						1							1				
546.	C. dichotomus,													2				
547.	C. baccifer,						1						1	1				+
545.	C. gemmifer,													1				+
844.	C. scaber,						1											
802.	C. flexuosus,						1											
802.	C. spicatus,						1											
844.	C. ebracteatus,						1	1										
803.	C. rugosus,						1		1									
540.	*Cordaicarpus.*																	
805.	C. stabilis,					1												
539.	C. Mansfieldi,								1					1				
804.	C. Gutbieri,								1	1				1				+
550.	C. ovatus,							1						1				+
551.	C. apiculatus,													1				
804.	C. costatus,							1	1			1		1				
804.	C. cinctus,								1									
805.	C. lineatus,					1								1				
551.	*Cordaistrobus.*																	
552.	C. Grand'Euryi,													1				
553.	*Dicranophyllum.*																	
553.	D. dichotomum,													1				
554.	D. dimorphum,													1				
556.	*Desmiophyllum.*																	
557.	D. gracile,													1				
557.	*Lepidoxylon.*																	
557.	L. anomalum,													1				
561.	*Cardiocarpus.*																	
562.	C. samaræformius,						1											
563.	C. Newberryi,					1												
563.	C. ingens,					1												
564.	C. affinis,					1												
564.	C. annulatus,						1		1									
565.	C. pachytesta,						1											
565.	C. (Ptilocarpus) bicornutus,														1			
567.	C. latus,						2							1				
567.	C. minus,					1	3											
567.	C. elongatus,						2											
813.	C. (Samaropsis) zonulatus,						1											
568.	C. (S.) late-alatus,					1	1											
569.	C. (S.) simplex,					1	1											
569.	C. orbicularis,					1	3											
570.	C. diminutivus,						1											
570.	C. fasciculatus,						1		1									
571.	C. apiculatus,						1							1				
810.	C. mamillatus,					1						1	1	1				
572.	C. regularis,						1							1				+
573.	C. congruens,						1							1				+
573.	C. marginatus,					1		1					1	1			1	+
813.	C. bicuspidatus,					2	1				1							+
597.	C. plicatus,							1										
597.	C. Trevortoni,							1										
810.	C. conglobatus,					1												
812.	C. crassus,					1	1											
809.	C. pachytesta,					1												
810.	C. ovalis,					1												
808.	C. Harveyi,					2												
808.	C. longicollis,					1								1				
807.	C. speciosus,					1												
806.	C. dilatatus,					1	1											
807.	C. patens,					1												

	1	2	3	4	5	6	7	8	9	10	11	12	13	14	15	16	17
811. C. divergens,							1	2									
811. C. latior,					1												
812. C. circularis,					1							1	1				
812. C. diplotesta,					1								1				
814. C. ellipticus,					1	1		2					1				
814. C. annularis,					1	1											
815. C. pusillus,					1	1		1									
574. *Rhabdocarpus*,													1				
575. R. insignis,								1									
575. R. Howardi,																1	
576. R. Jacksonianus,					1	2		3	1		1	3	1	1			
578. R. mulstriatus,					2	1	3	3	1			3	1				+
579. R. carinatus,						3											
579. R. acuminatus,						1		1									
579. R. lævis,						2									1		
580. R. Danai,						1							1				
581. R. clavatus,					4			1				1	1		1		+
844. R. Beinertianus,					1												+
581. R. amygdalæformis,					1	1	1					1				2	+
582. R. late-marginatus,					1												
583. R. minutus,					1	1											
583. R. cornutus,												1					
583. R. arcuatus,							1						1				
817. R. subglobosus,								1					1				
844. R. Bockschianus,					1			1					1				+
816. R. pachytesta,												1					
816. R. mamillatus,						1	1					1		1			
819. R. apiculatus,						1		1									
815. R. inflatus,													1				
818. R. emarginatus,													1				
818. R. abnormalis,													1				
818. R. tenax,													1				
815. R. late-costatus,					1	2		1									
584. *Trigonocarpus*.																	
584. T. Noeggerathi,						1						1					+
821. T. Dawesii,							1	1	1			1	1	1			+
586. T. Bertholetiformis,							1				1						
587. T. Saffordi,											1						
587. T. magnus,															1		
588. T. juglans,						2						1					
588. T. Hildreti,																1	
589. T. trilocularis,						1											
589. T. Parkinsoni,					3		1						1				+
590. T. olivæformis,					1	1		1				1		1			+
590. T. Menzelianus,												2					+
591. T. tricuspidatus,						1											
591. T. ornatus,						1											
592. T. multicarinatus,						1											
592. T. Giffordi,														1			
820. T. Adamsii,								2						1			
820. T. perpusillus,														1			
819. T. Schultzianus,								2						1			+
821. T. grandis,														1			
822. T. Kansaseanus,																1	
822. T. Starkianus,							2	1				1		1			
823. T. multistriatus,																1	
823. T. ampullæformis,					3			1									
590. T. subcylindricus,														1			
593. *Carpolithes*.																	
594. C. fasciculatus,															1		
595. C. cistula,												1					
595. C. corticosus,												1					
595. C. persicaria,												1					
596. C. acuminatus,						1	1										+

	1	2	3	4	5	6	7	8	9	10	11	12	13	14	15	16	17
596. C. retusus,	.	.	.	.	.	3	.	.	.	.	.	.	.	.	.	.	.
595. C. fragarioides,	.	.	.	.	.	1	.	.	.	.	.	1	.	.	.	.	+
598. C. spicatus,	1	.	.	.	.	.	.	.	.	.	.	.	.	.	.	.	.
598. C. lunatus,	1	.	.	.	.	.	.	.	.	.	.	.	.	.	.	.	.
598. C. siliqua,	1	.	.	.	.	.	.	.	.	.	.	.	.	.	.	.	.
826. C. latior,	.	.	.	.	.	1	.	.	.	.	.	.	.	.	.	.	.
825. C. perpusillus,	.	.	.	.	1	.	1	1	1	.	.	.	1	.	.	.	.
826. C. transsectus,	.	.	.	.	1	.	.	.	.	.	.	.	.	.	.	.	.
824. C. cerasiformis,	.	.	.	.	1	.	.	.	.	.	.	.	1	.	.	.	.
825. C. granularis,	.	.	.	.	.	.	.	1	.	.	.	.	.	.	.	1	.
824. C. Butlerianus,	.	.	.	.	.	.	.	1	.	.	.	.	.	.	.	.	.
824. C. conicus,	.	.	.	.	.	.	1	.	.	.	.	.	.	.	.	.	.
597. C. punctatus?	.	.	.	.	.	.	.	.	.	.	.	.	.	.	.	1	+
593. C. bifidus?	.	.	.	.	.	.	.	.	.	1	1	.	.	.	.	.	.
825. C. minimus,	.	.	.	.	.	.	.	1	.	.	.	.	1	.	.	+	.

REPORT OF PROGRESS P

EXPLANATIONS OF THE PLATES.

Plate A.

Figs. 1–6. *Taonurus marginatus*, Lx., p. 7.
Trans. Amer. Philos. Soc. vol. xiii, Pl. vii.
" 7. *T. . . Colletti*, Lx., p. 7.
" 8–8*b*. *Palæophycus Milleri*, Lx., p. 10.
Geol. Rept. of Indiana, 1875. Pl. i, f. 1, 1*a*, 1*b*.

Plate B.

Figs. 1, 2. *Conostichus Broadheadi*, Lx., p. 15.
" 3. *C. . . . . prolifer*, Lx., p. 16.
" 4. *C. . . . . ornatus*, Lx., p. 17.
Geol. Rept. of Inda., 1875, Pl. i, f. 6.
" 5, 6. *Asterophycus Coxii*, Lx., p. 12.
Ibid., Pl. ii, f. 1, 2.
" 7, 8. *A. . . . simplex*, Lx., p. 13.
" 9–10*a*. *Palæophycus gracilis*, Lx., p. 11.
Ibid., Pl. 1, f. 4–5*b*.
" 11, 11*a*. *Rhizomorpha Sigillariæ*, Lx., p. 3.
Proc. Am. Phil. Soc., No. 100, Pl. iv, f. 9, 9*b*.

Plate I.

Figs. 1. *Calamites cannæformis*, Schloth., p. 24.
" 2. *C. . . . ramosus*, Artis., p. 22.
" 3, 4. *C. . . . Suckowii*, Bt., p. 20.
" 5. *C. . . . approximatus*, Schloth., p. 26.
" 6. *C. . . . Cistii*, Brgt., p. 27.
" 7. *Bornia radiata*, Schp., p. 30.
Brgt. Hist. d. veg. foss., Pl. xxvi, f. 1, in parte.

Plate II.

Figs. 1–2*aa*. *Annularia longifolia*, Bt., p. 45.
" 2*b*, 2*bb*. *A.* . . . *inflata*, Lx., p. 47.
" 3, 3*a*. *Asterophyllites equisetiformis*, Bt., p. 35.
" 4–5*a*. *A.* . . . . *gracilis*, Lx., p. 42.
" 6, 7. *Sphenophyllum Schlotheimii*, Bt., p. 52.
" 8, 9. *Annularia sphenophylloides*, Zenk, p. 48.
" 10,10*a*. *Sphenophyllum bifurcatum*, Lx., p. 55.
Geol. Rept. of Arks., ii, Pl. i, f. 2, 2*a*.

Plate III.

Figs. 1-4. *Asterophyllites fasciculatus*, Lx., p. 41.
" 5–7. *A.* . . . . *ovalis*, Lx., p. 35.
" 8–10. Disposition of the fructifications of *Asterophyllites*, p. 43.
Flore carbon. de Grand 'Eury, Pl. vi, f. 4, 11.
" 11,12. *Asterophyllites*, fructified. (Original spec.), p. 44.
" 13. *Annularia sphenophylloides*, Zenk, var. *minor*, p. 48.
" 14. *Macrostachya infundibuliformis*, Schp., p. 60.
" 17–19*a*. Cones of *Macrostachya*, p. 828.
" 20. *Macrostachya aperta*, Lx., p. 829.
" 15,16. Sheaths of *Equisetites*, p. 62.
" 21 23*a*. *Trochophyllum clavatum*, Lx., p. 65, 794.
" 24-25*b*. *T.* " " *lineare*, Lx., p. 64, 791.

Plate IV.

Figs. 1, 1*a*. *Whittleseya elegans*, Newby, p. 523.
" 2. *Whittleseya integrifolia*, Lx., p. 524.
" 3. *W.* . . . *undulata*, Lx., p. 525.
" 4. *Neuropteris trichomanoides*, Bt., p. 79.
" 5. *N.* . . . *reniformis*, Bt., p. 77.

Plate V.

Figs. 1-6. *Neuropteris fimbriata*, Lx., p. 81.
F. 3, Geol. Rept. of Ills., vi, Pl. iv, f. 4; f. 6, Geol of Penna., 1858, Pl. iv, f. 17.
" 7, 8. *N.* . . . *dentata*, Lx., p. 82.
Ibid., Pl. iv, f. 21, Pl. v, f. 9.

PLATE VI.

Figs. 1–6. *Neuropteris gibbosa*, Lx., p. 84.
" 7–10. *N.* . . . *Rogersi*, Lx., p. 83.
" 11. *N.* . . . *auriculata?* Bt., p. 85.

PLATE VII.

Figs. 1, 1*a*. *Neuropteris anomala*, Lx., p. 118.
" 2–4*a*. *N.* . . . *inflata*, Lx., p. 86.
" 5–6*a*. *N.* . . . *Missouriensis*, Lx., p. 104.

PLATE VIII.

Figs. 1, 4, 5, 7, 9, 12. *Neuropteris hirsuta*, Lx., p. 88.
F. 9, 11. Geol. of Penna., 1858, Pl. iv, f. 4, 8.
" 2, 3, 6, 8, 10, 11. *N.* . . . *angustifolia*, Bt., p. 89.

PLATE IX.

Figs. 1–6. *Neuropteris Clarksoni*, Lx., p. 94

PLATE X.

Figs. 1–4. *Neuropteris plicata*, St., p. 96.
" 5–10. *N.* . . . *vermicularis*, Lx., p. 99.
F. 6, 8, Geol. Rept. of Ills., iv, Pl. vi, f. 1, 3.

PLATE XI.

Figs. 1–4. *Neuropteris Loschii*, Bt., p. 98.

PLATE XII.

Figs. 1 ?, 2–9. *Neuropteris tenuifolia*, Bt., p. 100.

PLATE XIII.

Figs. 1–3*a*. *Neuropteris Smithii*, Lx., p. 106.
" 4. *N.* . . . . . *Elrodi*, Lx., p. 107.
" 5, 6. *N.* . . . . *subfalcata*, Lx., p. 102.
" 7. *N.* . . . . . *biformis*, Lx., p. 121.
" 8. *N.* . . . . . *rotundifolia?* St., p. 97.
" 9, 9*a*. *N.* . . . . . *Grangeri*, Bt., p. 105.
" 10–12. *N.* . . . . . *aspera*, Lx., p. 121.

PLATE XIV.

Figs. 1–7. *Neuropteris Desorii*, Lx., p. 112.

PLATE XV.

Figs. 1. *Neuropteris Desorii*, Lx., p. 112.
" 2–5. *N. . . . . . rarinervis*, Bunb., p. 109

PLATE XVI.

Figs. 1–8. *Neuropteris callosa*, Lx., p. 115.
" 9–11*a*. *N. . . . . . crenulata*, Bt., p. 116

PLATE XVII.

Figs. 1–4. *Neuropteris Agassizi*, Lx., p. 117.

PLATE XVIII.

Figs. 1, 2. *Neuropteris Verbenæfolia*, Lx., p. 120
Geol. Rept. of Ills., iv, Pl. vi, f. 5, 6.
" 3–5. *N. . . . . Germari*, Goepp., p. 113.
F. 4, Geol. of Penna., 1858, Pl. xx, f. 5; f. 5, Ibid., Pl. v, f. 5.
" 6, 6*a*. *N. . . . . . coriacea*, Lx., p. 111.
Geol. Rept. of Ills., iv, Pl. viii, f. 7, 8.
" 7 *N. . . . . . Evenii*, Lx., p. 117.
Ibid., ii, Pl. xxxvi, f. 4.
" 8. *N. . . . . . pachyderma*, Lx.
Ibid., ii, p. 430, Pl. xxxvi, fig. 5.

PLATE XIX.

Figs. 1–5. *Odontopteris alpina*, Gein., p. 126.

PLATE XX.

Figs. 1–2. *Odontopteris Schlotheimii*, Bt., p. 136.

PLATE XXI.

Figs. 1. *Odontopteris alata*, Lx., p. 131.
Cat. Pottsville Sci. Assoc.. Pl. i, f. 1.
" 2. *O. . . . . . Brardii?* Bt., p. 132.
Geol. Rept. of Arks., ii, Pl. v, f. 7.
" 3, 4. *O. . . . . . sphenopteroides*, Lx., p. 139.
" 5, 6. *O. . . . . . subcrenulata*, Lx., p. 137.
" 7, 7*a*. *O. . . . . . abbreviata*, Lx., p. 138.
" 8. *O. . . . . . æqualis*, *Lx.*, p. 135.

Plate XXII.

Figs. 1, 1*a*. *Odontopteris Worthenii*, Lx., p. 130.
Geol. Rept. of Ills., ii, Pl. xxxvi, f. 1.
" 2, 3. *O. . . . . . tenuinervis*, Lx., p. 125.
Geol. of Penna., 1858, Pl. v, f. 7, 8.
" 4, 5. *O. . . . . . subcuneata*, Bunb., p. 134.
F. 5, Geol. Rept. of Ills., ii, Pl. xxxvi, f. 3.
" 6. *O. . . . . . heterophylla*, Lx., p. 129.
Ibid., Pl. xxxviii, f. 5.
" 7–9. *O. . . . . . cornuta*, Lx., p. 128

Plate XXIII.

Figs. 1–3. *Neuropteris capitata*, Lx., p. 103.
F. 1, Geol. Rept. of Ills., iv, Pl. vii, f. 1.
" 4–6. *Dictyopteris obliqua*, Bunb., p. 146.
" 7–10. *D. . . . . . rubella*, Lx., p. 145.
Ibid., Pl. vii, f. 2–5.
" 11. *Idiophyllum rotundifolium*, Lx., p. 160.

Plate XXIV.

Figs. 1. *Magalopteris Southwellii*, Lx., p. 148.
" 2, 2*a*. *M. . . . . fasciculata*, Lx., p. 150.
" 3. *M. . . . . . abbreviata*, Lx., p. 151.
" 4, 4*a*. *M. . . . . . marginata*, Lx., p. 152
" 5, 6. *Neuropteris fasciculata*, Lx., p. 93.
Geol. Rept. of Ills., iv, Pl. v, f. 2, 4.

Plate XXV.

Figs. 1–3. *Lesleya grandis*, Lx., p. 143.
" 4–6*a*. *Danæites macrophyllus* (Newb.), Lx., p. 187.
" 7, 7*a*. *Tæniopteris Smithii*, Lx., p. 153

Plate XXVI.

Figs. 1, 1*a*. *Lescuropteris Moorii* (Lx.), Schp., p. 162
Geol. of Penna., 1868, Pl. xix, f. 1.
" 4, 4*a*. *L. . . . . adiantites*, Lx., p. 163.
Ibid., Pl. xx, f. 1.
" 2–3. *Callipteridium Pardeei*, Lx., p. 169.

Plate XXVII.

Figs. 1–2. *Callipteridium Mansfieldi*, Lx. p. 166.
" 3, 3*a*. *C. . . . . . . neuropteroides*, Lx. p. 166
" 4–8*a*. *C. . . . . . . membranaceum*, Lx. p. 172

Plate XXVIII.

Figs. 1–3*a*. *Danæites Emersoni*, Lx. p. 157.
" 4–6*a*. *Alethopteris Gibsoni*, Lx. p. 183.
" 7, 7*a*. *A. . . . . . lonchitica*, Bt., p. 177.

Plate XXIX.

Figs. 1–5. *Alethopteris Serlii*, Bt., p. 176

Plate XXX.

Figs. 1–4. *Alethopteris Helenæ*, Lx. p. 179.
" 5, 5*a*. *Pecopteris angustissima ?* Bt., p. 257.

Plate XXXI.

Figs. 1–4. *Alethopteris ambigua*, Lx. p. 182.

Plate XXXII.

Figs. 1–7*a*. *Pseudopecopteris Mazoniana*, Lx. p. 190.
F. 2-7, Geol. Rept. of Ills., iv, Pl. ix, f. 3–7.

Plate XXXIII.

Figs. 1. *Alethopteris Gibsoni*, Lx., p. 183.
" 2–5. *Callipteridium inæquale*, Lx. p. 168.
" 6–7. *C. . . . . . . Owenii*, Lx., p. 167.
F. 6, Geol. Rept. of Arks., ii, Pl. ii, f. 1.
" 8–8*b*. *Pseudopecopteris Sheaferi*, Lx., p. 194.
" 9, 9*a*. *Callipteridium inflatum*, Lx., p. 174.

Plate XXXIV.

Figs. 1–3. *Pseudopecopteris nervosa* (Bt), Lx., p. 197.
" 4, 4*a*. *P. . . . . Pluckneti*, Schloth., p. 199.

Plate XXXV.

Figs. 6. *Pseudopecopteris dimorpha*, Lx., p. 201.
" 7, 7*a*. *P. . . . . Pluckneti*, Schloth., p. 199.

PLATE XXXVI.

Figs. 1–2*a*. *Callipteridium rugosum*, Lx., p. 169.

PLATE XXXVII.

Figs. 1, 1*a*. *Pseudopecopteris Newberryi*, Lx., p. 202.
" 2–2*b*. *P.* . . . . *muricata*, Bt., p. 203.
" 3. *P.* . . . . *Sillimanni*, Bt. p. 206.
" 4–5. *P.* . . . . *cordato-ovata* (Weiss), Lx., p. 205.
" 6, 6*a*. *P.* . . . . *acuta* (Bt.), Lx., p. 215.
" 7, 8. *P.* . . . . *subcrenulata*, Lx., p. 193.

PLATE XXXVIII.

Figs. 1–4. *Pseudopecopteris anceps*, Lx., p. 207.

PLATE XXXIX.

Figs. 1–3. *Callipteridium Aldrichi*, Lx., p. 171.
" 4–6*a*. *Pecopteris elliptica*, *Bunb.*, p. 245.
" 7, 8. *P.* . . . . (*Gon.*) *robusta*, Lx., p. 229.
" 9–10. *P.* . . . . (*Gon.*) *lanceolata*, Lx. p. 227.
Geol. Rept. of Ills., iv., Pl. xiii, f. 1–3.
" 11. *P.* . . . . (*Gon.*) *emarginata*, Goepp., p. 225.
Ibid., Pl. xiii, f. 4.
" 12–13*a*. *P.* . . . . *squamosa*, Lx., p. 235.
Ibid., Pl. xii, f. 1 in parte; Pl. xiii, f. 10, 11.
" 14–15*a*. *P.* . . . . *Strongii*, Lx., p. 236.
Ibid., Pl. xiii, f. 7–9.

PLATE XL.

Figs. 1–7*b*. *Pecopteris* (*Gon.*) *unita*, Brgt., p. 223.

PLATE XLI.

Figs. 1, 1*a*. *Pecopteris* (*Gon.*) *venulosa*, Lx., p. 230.
" 2–3*a*. *P.* . . . . (*Gon.*) *arguta*, Bt., p. 227.
" 4, 4*a*. *P.* . . . . *Cistii*, Bt., p. 243.
Hist. d. Veg. foss., Pl. cvi, f. 2, in parte.
" 5, 5*a*. *P.* . . . . *platyrachis*, Bt. p. 232.
Ibid., Pl. ciii, f. 5.

Figs. 6–7*b*. *P.* . . . . *arborescens*, Schloth., p 230.
" 8, 8*a*. *P.* . . . . *oreopteridis*, Bt., p. 238
Ibid., Pl. cv, f. 2, in parte.
" 9, 9*a*. *P.* . . . . *Miltoni*, Bt., p. 247.
" 10, 10*a*. *P.* . . . . *Clarkii*, Lx., p. 261.

PLATE XLII.

Figs. 1–5*a*. *Pecopteris Clintoni*, Lx., p. 251.

PLATE XLIII.

Figs. 1–7*a*. *Pecopteris vestita*, Lx., p.252.

PLATE XLIV.

Figs. 1, 1*a*, 3. *Pecopteris erosa*, Gutb., p. 255.
" 2, 2*a*. *P.* . . . . *cristata*, Gutb., p. 256.
" 4, 4*a*. *P.* . . . . *dentata*, Brgt., p. 240.

PLATE XLV.

Figs. 1–2*a*. *Pecopteris pennæformis*, Bt., p. 239.

PLATE XLVI.

Figs. 1–3*d*. *Pecopteris serpillifolia*, Lx., p. 237.
" 4–6*a*. *P.* . . . . *abbreviata?* Bt., p. 248.

PLATE XLVII.

Figs. 1–1*b*. *Oligocarpia Alabamensis*, Lx., p. 266.
" 2–2*b*. *Sphenopteris microcarpa*, Lx., p. 280.

PLATE XLVIII.

Figs. 1–3*b*. *Oligocarpia Gutbieri*, Goepp., p. 266.
F. 3*a* and 3*b*, Geinitz, Verst., Pl. xxxiii, f. 7*a*, 7*b*.
" 4–5*b*. *Pecopteris lyratifolia*, Goepp., p. 259.
" 6–6*b*. *P.* . . . . *solida*, Lx., p. 261.
Geol. Rept. of Ills., iv, Pl. xi, f. 5–7.
" 7–7*b*. *P.* . . . . *stellata*, Lx., p. 260.
Ibid., ii, Pl. xxxvii, f. 4–4*b*.
" 8–8*b*. *Sorocladus stellatus*, Lx., p. 328.
Geol. Rept. of Arks., ii, Pl. ii, f. 2–2*b*.

Figs. 9–9*b*. *S.* . . . *asteroides*, Lx., p. 328.
Geol. Rept. of Ills., iv, Pl. xiv, f. 6–7*b*.

" 10–10*b*. *S.* . . . *sagittatus*, Lx., p. 329.
Ibid., Pl. xiv, f. 3–5.

" 11. *S.* . . . . *ophioglossoides*, Lx., p. 329.

PLATE XLIX.

Figs. 1–3. *Archæopteris Bockschiana?* Goepp., 306.
Geol. of Penna., 1858, Pl. iv, f. 1–1c.

" 5. *A.* . . . . . . *minor*, Lx., p. 302.

" 6, 7. *A.* . . . . . . *obtusa*, Lx., p. 301.
F. 6, ibid., Pl. i, f. 11.

" 8, 8*a*. *A.* . . . . . . *stricta*, Andrews, p. 303.
Geol. Rept. of Ohio, ii, Pl. xiix, f. 2, in parte.

" 9, 9*a*. *A.* . . . . . . *Alleghaniensis*, Meek, p. 307.
Bull. Phil. Soc. of Washington, Dec. 1875, Pl. 1, f. 2.

PLATE L.

Figs. 1–4. *Archæopteris minor*, Lx., p. 302.

" 5. *A.* . . . . . . *Hybernica*, Forb., p. 305.

" 6–6*c*. *Triphyllopteris Lescuriana*, Meek, p. 297.
Phil. Soc. Wash., 1875, Pl. ii.

" 7. *Rhacophyllum truncatum*, Lx., p. 311.

PLATE LI.

Figs. 1–1*b*. *Pseudopecopteris speciosa*, Lx., p. 216.

PLATE LII.

Figs. 1–3*b*, 8. *Pseudopecopteris irregularis*, St., p. 211.

" 4, 4*a*. *P.* . . . . *latifolia*, Bt., p. 215.

" 9–10*a*. *P.* . . . . *decipiens*, Lx., p. 214.

" 6–7*a*. *Eremopteris microphylla*, Lx., p. 296.

PLATE LIII.

Figs. 1, 2. *Eremopteris crenulata*, Lx., p. 292.

" 3. *E.* . . . . . *flexuosa*, Lx., p. 293.

" 4. *E.* . . . . . *dissecta*, Lx., p. 293.

" 5–6. *E.* . . . . . *artemisiæfolia*, Bt., p. 293.

" 7–7*a*. *E.* . . . . . *elegans*, Ett., p. 294.

" 8, 8*a*. *E.* . . . . . *Missouriensis*, Lx., p. 295.

PLATE LIV.

Figs. 1–3*a*. *Sphenopteris mixta*, Schp., p. 276.

PLATE LV.

Figs. 1, 1*a*. *Sphenopteris subalata?* Gein., p. 272.
" 2–2*b*. *S.* . . . . *Brittsii*, Lx., p. 277.
" 3–4*a*. *S.* . . . . . . *goniopteroides*, Lx., p. 269.
" 5, 5*a*. *S.* . . . . . . *Hœninghausii*, Bt., p. 288.
" 6, 6*a*. *S.* . . . . . . *elegans*, Bt., p. 287.
" 7, 7*a*. *S.* . . . . . . *Larischi*, Stur., p. 288.
" 8–9*a*. *S.* . . . . . . *tridactylites*, Bt. p. 284.

PLATE LVI.

Figs. 1, 1*a*. *Pseudopecopteris spinulosa*, Lx., p. 195.
Geol. Rept. of Ills., iv, Pl. xi, f. 1, 2.
" 2–3*b*. *P.* . . . . *hymenophylloides*, Lx., p. 196.
Ibid., Pl. x, f. 1–4.
" 4, 4*a*. *Sphenopteris splendens*, Lx., p. 282.
Ibid., Pl. xix, f. 2.
" 5–5*b*. *Sphenophyllum cornutum*, Lx., p. 56.
Same specimens as f. 4.

PLATE LVII.

Figs. 1, 1*a*. *Rhacophyllum flabellatum?* St., p. 311.
" 2. *R.* . . . . *hirsutum*, Lx., p. 318.
" 3, 3*a*. *R.* . . . . . . *cornutum*, Lx., p. 317.
" 4, 4*a*. *R.* . . . . *corallinum*, Lx., p. 317
" 5. *R.* . . . . *Clarkii*, Lx., p. 319.
Geol. Rept. of Ills., iv, Pl. xvi, f. 2.
" 6. *R.* . . . . *expansum*, Lx., p. 323.
" 7–8. *R.* . . . . *inflatum*, Lx., p. 323.
Ibid., Pl. xvi, f. 6–6*a*.
" 9–11. *R.* . . . . . . *adnascens*, Ll. & H., p. 321.
F. 9–10, Ibid., Pl. xvi, f. 7, 8.

PLATE LVIII.

Figs. 1, 2. *Rhacophyllum membranaceum*, Lx., p. 312.
" 3. *R.* . . . . *hamulosum*, Lx., p. 321
" 4, 5. *R.* . . . . *spinosum*, Lx., p. 320.
" 6, 7. *R.* . . . . *fucoideum*, Lx., p. 325.

PLATE LIX.

Figs. 1. *Stemmatopteris hirsuta*, Lx., p. 337.
" 2. *S. . . . . . squamosa*, Lx., p. 339.
" 3. *S. . . . . . . punctata*, Lx., p. 339.
Geol. of Penna., 1858, Pl. xiii, f. 1.
" 4. *S. . . . . . . mimica*, Lx., p. 341.
" 5. *S. . . . . . . angustata*, Lx., p. 339.
" 6. *S. . . . . . . polita*, Lx., p, 342.
" 7. *S. . . . . . . insignis*, Lx., p. 340.
Geol. Rept. of Ills., ii, Pl. xlix, f. 1.
" o. *Caulopteris obtecta*, Lx., p. 344.
Ibid.. iv, Pl. xxviii, f. 1.

PLATE LX.

Figs. 1, 2. *Caulopteris Giffordi*, Lx., p. 343.
" 3. *C. . . . . . Mansfieldi*, Lx., p. 346.

PLATE LXI.

Figs. 1, 2. *Halonia tortuosa*, Ll. & H., p. 413.
" 3, 3*a*. *H. . . . flexuosa*, Lx., p. 416.
" 4. *Megaphitum Goldenbergii*, Weiss., p. 349.
" 5. *Halonia pulchella*, Lx., p. 417.
Geol. Rept. of Aks., ii, Pl. iii, f. 3 in parte.

PLATE LXII.

Figs. 1, 1*a*. *Lycopodites Meekii*, Lx., p. 357.
Geol. Rept. of Ills., iv, Pl. xxvi, f. 6, 6*a*.
" 2, 2*a*. *L. . . . . . pendulus*, Lx., p. 357.
" 3–3*d*. *Lepidodendron squamiferum*, Lx., p. 376
" 4, 4*a*. *L. . . . . . . rhombicum*, St., p. 382.
" 5. *L. . . . . . . cyclostigma*, Lx., p. 394.
" 6–8*a*. *L. . . . . . . Weltheimianum*, St., p. 374.

PLATE LXIII.

Figs. 1–2. *Lepidodendron Brittsii*, Lx., p 368.
" 3–5*a*. *L. . . . . . . lanceolatum*, Lx., p. 369.
" 6–6*c*. *L. . . . . . . scutatum*, Lx., p. 369.
" 7–8. *L. . . . . . . latifolium*, Lx., p. 370.
" 9–10*a*. *L. . . . . . . forulatum*, Lx., p. 390.
Geol. Rept. of Ills., iv, Pl. xxiii, f. 5–8.

Plate LXIV.

Figs. 1. *Lepidodendron, aculeatum*, St., p. 371.
" 2. *L.* . . . . *diplotegioides*, Lx., p. 390.
Geol. Rept. of Ills., ii, Pl. xlix, f. 2.
" 3. *L.* . . . . *obovatum*, St., p. 384.
" 4. *L.* . . . . *costatum*, Lx., p. 381.
Ibid., Pl. xliv, f. 7.
5. *L.* . . . . *turbinatum*, Lx., p. 382.
Ibid., Pl. xliv, f. 6.
" 6. *L.* . . . . *Andrewsii*, Lx., p. 389.
" 7. *L.* . . . . *cuspidatum*, Lx., p. 388.
" 8, 9. *L.* . . . . *Worthenii*, Lx., p. 388.
Ibid., Pl. xlix, f. 4. 5.
" 10. *L.* . . . . *distans*, Lx., p. 387.
Geol. of Penna., 1858, Pl., xvi, f. 4.
" 11. *L.* . . . . *rimosum*, St., p. 392.
" 12. *L.* . . . . *Mielicki*, Goepp., p. 395.
" 13, 14. *L.* . . . . *modulatum*, Lx., p. 385.
Geol. Rept. of Arks., ii, Pl. iii, f. 1, 1*a*.
" 15. *L.* . . . . *vestitum*, Lx., p. 379.
" 16–18. *L.* . . . . *clypeatum*, Lx., p. 380.

Plate LXV.

Figs. 1. *Ulodendron elongatum*, Lx., p. 405
Geol. Rep,. of Ill., iv, Pl. xxiii, f. 4.
" 2–4. *U.* . . . . . *ellipticum*, St., p. 404.
Ibid., Pl. xxii, f. 3, xxiii, 1–3.
" 5, 5*a*. *U.* . . . . . *punctatum*, St., p. 405.

Plate LXVI.

Figs. 1. *Ulodendron minus*, Ll. & H., p. 403.
" 2, 2*a*. *U.* . . . . . *commutatum*, Schp., p. 401.
" 3–3*a*. *U.* . . . . . *majus*, Ll. & H., p. 401.
Geol. Rept. of Ills., iv, Pl. xxii, f. 4*a*, *b*. *c*.

Plate LXVII.

Figs. 1, 1*a*. *Halonia secreta*, Lx., p. 417.
" 2, 2*a*. *Ulodendron Mansfieldi*, Lx., p. 414.
" 3. *Dechenia striata*, Lx., p. 431.

Plate LXVIII.

Figs. 1. Fragment of cone of *Lepidophloios laricinus*, St., p. 428.
Roehl. Paleont., xviii, Pl. xiii, f. 1*a*.

" 2. *Lepidophloios macrolepidotus*, Gold., p. 424.

" 3. Cone of *Lepidophloios auriculatus*, Lx., p. 421.
Geol. Rept. of Ills., iv, Pl. xxx, f. 1.

" 4. Scale of the same.

" 5. *Lepidophyllum auriculatum*, Lx., p. 421.
Ibid., Pl. xxiv, f. 1 in parte.

" 6. *Lepidophloios ;* fragment of a cone with scales partly covering macrospores, p. 428.

" 7–7*b*. Same; agglomerations of macrospores, p. 428.

" 8, 8*a*. *L. . . . ? Sigillarioides*, Lx., p. 425.

" 9, 9*a*. *L. . . . . protuberans*, Lx., p. 425.
Ibid., Pl. xxvi, f. 1, 2.

" 10. *L. . . ichthyoderma*, Lx., p. 426.

Plate LXIX.

Figs. 1–2. *Lepidostrobus*, (*macrocystis*,) *Salisburyi*, Lx., p. 443.

" 3. *Lepidocystis pectinatus*, Lx., p. 454.

" 4. *L. . . . . . lineatus*, Lx., p. 454.

" 5. *L. . . . . . quadrangularis*, Lx., p. 455

" 6, 7. *L. . . . . obtusus*, Lx., p. 455.

" 8. *Lepidophyllum foliaceum*, Lx., 445.

" 9, 10. *L. . . . . . . truncatum*, Lx., p. 458.

" 11–14. Agglomeration of macrospores (*Sporocystis*,) p. 458.

" 15, 15*a*. *Sporocystis planus*, Lx., p. 458.

" 16–17*a*. *Lepidocystis angularis* Lx., p. 456.

" 18–20. *L. . . . . . vesicular* , Lx., p. 457.

" 21–23. *L. . . . . . fraxiniformis*, Lx., p. 457.

" 24, 24*a*. *L. . . . . . bullatus*, Lx., p. 458.

" 25, 25*a*. *Lepidostrobus incertus*, Lx., p. 442.

" 26. *L. . . . . . variabilis*, Ll. & H., p. 434

" 27, 28. *L. . . . . . hastatus*, Lx., p. 438.

" 29. *L. . . . . . oblongifolius*, Lx., 437.
Geol. Rept. of Ills., iv, Pl. xxx, f. 3.

Figs. 30, 30*a*. *Lepidostrobus lancifolius*, Lx., 436.
Ibid., Pl. xxxi, f. 7, 7, *b*.
" 31. *Lepidophyllum affine*, Lx., p. 447.
" 32. *L.* . . . . . . *ovatifolium*, Lx., p. 438.
Ibid., Pl. xxx, f. 2*b*.
" 33. *L.* . . . . . . *brevifolium*, Lx., p. 453.
Geol. of Penna., 1858, Pl. xvii, f. 6.
" 34. *Lepidophyllum Mansfieldi*, Lx., p. 449.
" 35. *L.* . . . . . . *rostellatum*, Lx., p. 451
Geol. Rept. of Ills., iv, Pl. xxxi, f. 8.
" 36. *L.* . . . . . . *striatum*, Lx., p. 452.
" 37. *L.* . . . . *acuminatum*, Lx., p. 450
" 38. *L.* . . . . *lanceolatum*, Bt., p. 436.
" 39. *L.* . . . . *linearifolium*, Lx., p. 452
" 40, 41. *L.* . . . . *Morisianum*, Lx., p. 448.

PLATE LXX.

Figs. 1–1*b*. *Syringodendron Porteri*, Lx., p. 502.
Geol. Rept. of Ills., iv, Pl. xxvii, f. 4–6.
" 2, 2*a*. *S.* . . . . *pachyderma*, Bt., p. 503.
" 3–3b. *S.* . . . . *Brongniarti*, St., p. 504.
" 4, 4*a*. *S.* . . . . *cyclostigma*, Bt., p. 505.
" 5–9. *Sigillaria reniformis*, Bt., p. 501.

PLATE LXXI.

Figs. 1–3. *Sigillaria lævigata*, Bt., p. 500.
" 4. *S.* . . . . *Pittstoniana*, Lx., p. 493.
" 5. *S.* . . . . *marginata*, Bt., p. 498.
" 6. *S.* . . . . *Sillimanni*, Bt., p. 493.
" 7, 8. *S.* . . . . *ovalis*, Lx., p. 495.

PLATE LXXII.

Figs. 1. *Sigillaria hexagona*, Bt., p. 483.
" 2–4*b*. *S.* . . . . *tessellata*, Bt., p. 481.
" 5, 6. *S.* . . . . *mamillaris*, Bt., p. 483.
" 7, 8. *S.* . . . . *Lescurii*, Schp., p. 485.
Potts. Cat., Pl. ii, f. 1, 2.
" 9. *S.* . . . . *attenuata*, Lx., p. 488.
Ibid., Pl. ii, f. 3.
" 10. *S.* . . . . *leptoderma*, Lx., p. 489.
" 11. *S.* . . . . *Wolzii*, Bt., p. 492.
" 12–12*b*. *S.* . . . . *Lacoei*, Lx., p. 499.

PLATE LXXIII.

Figs. 1. *Sigillaria polita*, Lx., p. 490
Geol. of Penna., 1858, Pl. xiv, f. 3.

" 2. *S.* . . . . *Yardleyi*, Lx., p. 491.
Potts. Cat., Pl. ii, f. 4.

" 3–6. *S.* . . . . *monostigma*, Lx., p. 468.
F. 6, Geol. Rept. of Ills., iv, Pl. xxvi, f. 5.

" 7. *S.* . . . . *ichthyolepis*, St., p. 482.

" 8–16*a*. *S.* . . . . *Brardii*, Bt., p. 477.

" 17. *S.* . . . . *fissa*, Lx., p. 470.

" 18. *S.* . . . *obliqua*, Bt. p. 470.

" 19,19*a*. *S.* . . . . *reticulata*, Lx., p. 473

" 20. *S.* . . . . *stellata*, Lx., p. 474.
Geol. of Penna., 1858, Pl. xiv, f. 2, in parte.

" 21. *S.* . . . . *Schimperi*, Lx., p. 474.
Ibid., Pl. xiv, f. 1, in parte.

PLATE LXXIV.

Figs. 1. *Stigmaria ficoides*, Bt., p. 514. Size reduced.

" 2. *S.* . . . . *ficoides*, var. Goepp., p. 514.
Fragment of surface, natural size.
Goepp. Gatt., Pl. ix, f. 8.

" 3. *S.* . . . . *ficoides* var. *undulata*, Goepp., p. 515.
Ibid., Pl. ix, f. 9.

" 4. *S.* . . . . *ficoides* var. *stellata*, Goepp., p. 515.
Ibid., Pl. x, f. 12, in parte.

" 6. *S.* . . . . *ficoides* var. *minus*, auct. not described.

" 8. *S.* . . . . *umbonata*, Lx., p. 516.
Ibid., Pl. ix, f. 7.

" 5, 7. *S.* . . . . *stellaris*, Lx., p. 516.
F. 7, Geol. Rept. of Ills., iv, Pl. xxix, f. 3; f. 5, decorticated origical spec.

" 9. *S. Halonia tuberculata?* Bt., p. 411.
Ibid., Pl. xxix, f. 1.

" 10–10*b*. Trunk of *Sigillaria Owenii*, Lx., p. 507.
With scars of the stem 10*a* and of the root 10*b*.

Figs. 11. *Stigmaria* (Trunk, size reduced), p. 509, etc.
Schimp. Veget. Paleont., Pl. lxix, f. 2.

Figs. 12–13. Stigmaria (leaves and tubercles), p. 509, etc.
" 14, 15. *Knorria imbricata*, St., p. 407.

PLATE LXXV.

Figs. 1. *Stigmarioides Evenii*, Lx., p. 333.
Geol. Rept. of Ills., ii, Pl. xxxix, f. 9.
" 2. *S.* . . . . *truncatus*, Lx., p. 334
Ibid., iv, Pl. xxix, f. 4.
" 3. S. . . . . *villosus*, Lx., p. 334.
Ibid., Pl. xxxi, f. 1.
" 4. *S.* . . . . . *tuberosus*, Lx., p. 335.
Ibid., Pl. xxix, f. 5.
" 5. *S.* . . . . *linearis*, Lx., p. 335.
Ibid., Pl. xxxi, f. 2.
" 6. *Sigillarioides radicans*, Lx., p. 517.
Ibid., Pl. xxxi, f. 4.
" 7. *Rachiopteris affinis*, Lx., p. 331.
Ibid., Pl. xxvii, f. 9.
" 8. *R.* . . . . *selago*, Lx., p. 332.
Ibid., Pl. xxxi, f. 3.
" 9. *Rhizolites palmatifidus*, Lx., p. 518.
Geol. Rept. of Arks., ii, Pl. v, f. 9.
" 10–10*b*. *Pachypteris gracillima*, Lx., p. 308.
Geol. Rept. of Ills., iv, Pl. xix, f, 6–8.
" 11. *Spirangium multiplicatum*, Lx., p. 520.
" 12. *S.* . . . . *appendiculatum*, Lx., p. 520.
Ibid., Pl. xxvii, f. 11.
" 13–15. *S.* . . . *Prendelii*, Lx., p. 519.
Ibid., Pl. xxvii, f. 10, 12, 13.
" 16. *Calamodendron approximatum?* Cotta, p 33.
" 17. *Calamites? gracilis*, Lx., p. 29.

PLATE LXXVI.

Figs. 1–2*a*. *Cordaites validus*, Lx., p. 529.
" 3–8*b*. *C.* . . . *borassifolius*, Ung., p. 532.
" 4–4*b*. *C.* . . . *Mansfieldi*, Lx. (Stem and flowers,) p. 537.
" 5–6*b*. *Cordaianthus gemmifer*, Gr. d'H., p. 545.

PLATE LXXVII.

Figs. 1–2*a*. *Cordaites grandifolius*, Lx., p. 530.
" 3, 3*a*. *C.* . . . *diversifolius*, Lx., p. 535
" 4–4*b*. *C.* . . . *gracilis*, Lx., p. 539

PLATE LXXVIII.

Figs. 1–2. *Cordaites Mansfieldi*, Lx., p. 537.

PLATE LXXIX.

Figs. 1–4. *Cordaites serpens*, Lx., p. 542.

PLATE LXXX.

Figs. 1–3. *Cordaites costatus*, Lx., p. 540.
" 4. *Tæniophyllum decurrens*, Lx., p. 464.

PLATE LXXXI.

Figs. 1. *Tæniophyllum decurrens*, Lx., p. 464.
" Stem of *Cordaites* with axis of *Artisia*, p. 528.

PLATE LXXXII.

Figs. 1. *Desmiophyllum gracile*, Lx., p. 557.
" 2, 2*a*. *Tæniophyllum contextum*, Lx., p. 465.
" 3–4*a*. *Cordaistrobus Grand' Euryi*, Lx., p. 552.

PLATE LXXXIII.

Figs. 1–3. *Dicranophyllum dimorphum*, Lx., p. 554.
" 4. *Tæniophyllum deflexum*, Lx., p. 465.
" 5. *Lepidoxylon anomalum*, Lx., p. 557.
" 6, 6*a*. *Cordaicarpus apiculatus*, Lx., p. 551.
" 7. *C. . . . . . . ovatus*, Grd'H., p. 550.
" 8–11. *C. . . . . . . Gutbieri*, Grd'H., p. 549.

PLATE LXXXIV.

Figs. 1. *Lepidoxylon anomalum*, Lx., p. 557.

PLATE LXXXV.

Figs. 1. *Trigonocarpus Noeggerathii*, Ll. & H., p. 584.
Geol. Rept. of Ills., iv. xxxi, f. 16.
" 2, 3, 25. *T. . . . . . . Dawesii*, Ll. & H., p. 586.
" 4. *T. . . . . . . juglans*, Lx., p. 588.
Ibid. ii, Pl. xivi, f. 3.
" 5, 6. *T. . . . . . . Giffordi*, Lx., p. 592.
Figs. 7–8*a*. *T. . . . . . . olivæformis*, Ll. & H., p. 590.
" 9, 10. *T. . . . . . . subcylindricus*, Lx., p. 590.
" 11. *T. . . . . . . Menzelianus*, Goepp., p. 590

Figs. 12, 13. *Trigonocarpus ornatus*, Newb., p. **591.**
Geol. Rept. of Ohio, i, Pl. xlii, f. 7, 7*a*.
" 14. *Rhabdocarpus clavatus*, St., p. 581.
Geol. Rept. of Ills., iv, Pl. xxxi, f. 11.
" 15. *Rhabdocarpus cornutus*, Lx., p. 583.
" 16. *Carpolithes bifidus?* Lx., p. 593.
" 17–19. *C. . . . . . Jacksonianus*, Lx., p. 576.
" 20. *Rhabdocarpus clavatus?* Gein., p. 581.
" 21. *Cordaicarpus Mansfieldi*, Lx., p. 539.
" 22, 23. *Rhabdocarpus multistriatus?* St., p. 578.
" 24. *R. . . . . Howardi*, Lx., p. 575.
" 26. *R. . . . . insignis*, Lx., p. 575.
" 27, 28. *R. . . . . amygdalæformis*, Goepp., p. 581.
" 29. *R. . . . . latemarginatus*, Lx., p. 582.
" 30, 30*a*. *Carpolithes fasciculatus*, Lx., p. 570.
" 31, 31*a*. *Cardiocarpus regularis?* St., p. 572.
" 32–33*a*. *C. . . . . . . mamillatus*, Lx., p. 571.
Fig. 33, 33*a*, Geol. Rept. of Ills., iv, Pl. xxxi, f. 12, 13.
" 34, 35. *C. . . . . . . ingens*, Lx., p. 563.
Geol. Rept. of Arks., ii, Pl. iv, f. 4, 4*a*.
" 36, 37. *C. . . . . . . annulatus?* Newb., p. 564.
" 38. *C. . . . . . . minor*, Newb., p. 548.
Geol. Rept. of Ohio, i, Pl. xliii, f. 4.
" 39. *C. . . . . . . affinis*, Lx., p. 564.
Geol. Rept. of Arks., ii, Pl. iv, f. 5.
" 40. *C. . . . . . . orbicularis*, Newb., p. 569.
Geol. Rept. of Ohio, ii, Pl. xliii, f. 10.
" 41. *C. . . . . . . elongatus*, Newb., p. 567.
Ibid., Pl., xliii, f. 5.
" 42, 43. *C. . . . . . . bicuspidatus*, St., p. 573.
Ibid., Pl. xliii, f. 9, 9*a*.
" 44, 45. *C. . . . . . . zonulatus*, Lx., p. 568.
" 46, 47. *C. . . . . . . late-alatus*, Lx., p. 568.
" 48, 50. *C. . . . . . . simplex*, Lx., p. 569.
" 51, 51*a* *C. (Ptilocarpus) bicornutus*, Lx., p. 565.
" *52. *Rhabdocarpus arcuatus*, Lx. p. 583.

* Plates LXXXVI, LXXXVII are intercalated in the text of Vol. II, pp. 544 and 560.

PLATE LXXXVIII.

Fig. 1. Dendrophycus Desorii, Lx., p. 699.

PLATE LXXXIX.

Figs. 1, 2. Annularia tuberculata, Lx., p. 723.
" 3, 4. Calamostachys ovalis, Lx., p. 717.
" 5, 5a. C. " brevifolia, Lx., p. 718.

PLATE XC.

Fig. 1. Volkmannia crassa, Lx., p. 719.
" 2. V. " prælonga, Lx., p. 720.
" 3. Macrostachya spike, p. 721.
" 4. Volkmannia fertilis, Lx., p. 720.
' 5. Equisetites gracilis, Lx., p. 729.

PLATE XCI.

Figs. 1, 2. Calamostachys lanceolata, Lx., p. 715.
" 3. Macrostachya with pedicel, p. 721.
" 4, 4a. Calamites ramifer, ? Stur, p. 703.
" 5. Bornia radiata, Schp.,? p. 706.
" 6 Sphenophyllum longifolium, Germ., p. 726.

PLATE XCII.

Figs. 1, 4. Calamites ramosus, Artis, p. 702.
" 5. Calamodendron, Brgt., p. 708.
" 6, 6a. Annularia Emersoni, Lx., p. 725.
" 7, 7a. A. " cuspidata, Lx., p. 725.
" 8, 8a. A. " minuta,? Brgt., p. 725.
" 9–10a. Sphenophyllum tenerrimum, Ett., p. 728.
" 11-11c. Didymophyllum (Sigillaria) Owenii, Lx., p. 801.

PLATE XCIII.

Fig. 1. Calamostachys lanceolata, Lx., p. 715.
" 2. Bornia radiata,? Brgt., p. 707.
" 3–7. Asterophyllites gracilis, Lx., p. 714.
" 8. Sphenophyllum angustifolium, Germ., p. 726.
" 9–9a. S. " saxifragæfolium, Germ., p. 726.

PLATE XCIV.

Figs. 1, 2. Neuropteris decipiens, Lx., p. 733.
" 3. N. " oblongifolia, Lx., p. 732.
" 4–7. N. " Carrii, Lx., p. 731.
" 8. Tæniopteris truncata, Lx., p. 743.

PLATE XCV.

Fig. 1, 1 a. Neuropteris Blissii, Lx., p. 737.
" 2. N. " oblongifolia, Lx., p. 732.
" 3–8. N. " Griffithii, Lx., p. 737.

PLATE XCVI.

Figs. 1, 2. Neuropteris Elrodi, Lx., p. 735.
" 3, 3 a. N. " Smithii, Lx., p. 734.
" 4–7. N. " aspera, Lx., p. 736.

PLATE XCVII.

Figs. 1–3. Odontopteris monstruosa, Lx., p. 741.
" 4. O. " affinis, Lx., p. 742.
" 5. O. " deformata, Lx., p. 743.
" 6. O. " æqualis, Lx., p. 743.
" 7, 7 a. O. " patens, Lx., p. 740.
" 8. O. " neuropteroides, Roem., p. 740.
" 9. Megalopteris rectinervis, Lx., p. 744.

PLATE XCVIII.

Fig. 1. Callipteridium inæquale, Lx., p. 745.
" 2. Pseudopecopteris macilenta, Ll. & H., p. 755.
" 3. P. " hispida, Lx., p. 755.
" 4, 4 a. P. " dimorpha, Lx., p. 751.
" 5, 5 a. Alethopteris crassa, Lx., p. 748.
" 6, 6 a. Pecopteris Georgiana, Lx., p. 759.

PLATE XCIX.

Figs. 1–2 a. Callipteridium rigidum, Lx., p. 746.
" 3. C. " sinuatum?, Lx., p. 745.
" 4, 4 a. Pecopteris robusta, Lx., p. 756.

Plate C.

Figs. 1, 2. Pecopteris quadratifolia, Lx., p. 757.
" 3, 3 a. P. " Carrii, Lx., p. 759.
" 4, 5. Sorocladus sagittatus, Lx., p. 761.

Plate CI.

Figs. 1, 1 b. Sphenopteris Gravenhorstii, var. B, Brgt., p. 763.
" 2. S. " (Diplothmema) Tracyana, Lx., p. 767.
" 3. S. " solida, Lx., p. 769.
" 4–8 a. Pecopteris serrulata, Hartt, p. 757.

Plate CII.

Figs. 1, 1a. Sphenopteris cristata, St., p. 761.
" 2. S. " hymenophylloides, Brgt., p. 764.
" 3–4a. S. " Brittsii, Lx., p. 765.

Plate CIII.

Figs. 1–3. Pseudopecopteris nummularia, Gutb., p. 752.
" 4–5 a. Sphenopteris inæquilateralis, Lx., p. 765.
" 6. S. " obovata, Ll. & H., p. 770.
" 7–7 b. S. " Harveyi, Lx., p. 766.

Plate CIV.

Figs. 1, 1a. Sphenopteris communis, Lx., p. 762.
" 2–4. Eremopteris Cheathami, Lx., p. 770.
" 5, 5 a. Sphenopteris cristata, var. minor, Lx., p. 762.
" 6. S. " divaricata, Goepp., p. 767.
" 7–10. S. " Royi, Lx., p. 768.

Plate CV.

Figs. 1–4. Lepidophloios dilatatus, Lx., p. 781.

Plate CVI.

Figs. 1. Lycopodites arborescens, Lx., p. 778.
" 2. L. " simplex, Lx., p. 779.
" 3, 4. L. " flexifolius, Lx., p. 779.

Plate CVII.

Fig. 1. Lycopodites Lacoei, Lx.,p. 780.
" 2. Lepidodendron Sternbergii, Brgt., &
" " Lepidostrobus variabilis, Ll. & H., p. 783.
" 3. Lepidodendron marginatum, Presl., p. 784.
" 4, 5. Lepidophyllum fallax, Lx., p. 786.
" 6, 7. L. " Campbellianum, Lx., p. 786.
" 8. L " gracile, Lx., p. 786.
" 9. L. " minutum, Lx., p. 787.
" 10. L. " coriaceum, Lx., p. 787.
" 11. L. " elegans, Lx., p. 787.
" 12. L. " tumidum, Lx., p. 788.
" 13, 14. L. " cultriforme, Lx., p. 785.
" 15. Sigillaria Williamsii, Lx., p. 801.

Plate CVIII.

Fig. 1. Lepidostrobus truncatus, Lx., p. 784.
" 2. Lepidophyllum cultriforme, Lx., p. 785.
" 3. Taeniophyllum brevifolium, Lx., p. 788.
" 4, 5. Sigillaria Leverettii, Lx. p. 800.
" 6. S. " mamillaris, Brgt., p. 799.

Plate CIX.

Fig. 1. Cordaianthus spicatus, Lx., p. 802.
" 2. C. " flexuosus, Lx., p. 802.
" 3. Cordaicarpus Gutbieri, Grd'E., p. 804.
" 4. C. " costatus, Lx., p. 804.
" 5, 6. C. " cinctus, Lx., p. 804.
" 7. Cardiocarpus mamillatus,? Lx., p. 810.
" 8, 9. C. " ovalis, Lx., p. 810.
" 10, 11. C. " conglobatus, Lx. p. 810.
" 12. C. " crassus, Lx., p. 812.
" 13–15. C. " pachytesta, Lx., p. 809.
" 16, 17. C. " elongatus, Ny., p. 809.
" 18–21, Trigonocarpus ampullæformis, Lx., p. 823.
" 22, 23. Cardiocarpus Harveyi, Lx., p. 808.
" 24, 25. C. " longicollis, Lx., p. 808.

PLATE CX.

Fig.	1.	Cardiocarpus speciosus, Lx., p. 807.
"	2.	C. " dilatatus, Lx., p. 806.
"	3.	C. " patens, Lx., p. 807.
"	4.	C. " divergens, Lx., p. 811.
"	5, 11, 12.	C. " latior, Lq., p. 811.
"	6–9.	C. " crassus, Lx., p. 812.
"	10.	C. " circularis, Lx., p. 812.
"	13.	C. " diplotesta, Lx., p. 812.
"	14–17.	C. " zonulatus, Lx., p. 813.
"	18–22.	C. " bicuspidatus, St., p. 813.
"	23–27.	C. " ellipticus, St., p. 814.
"	28–30.	C. " annularis, St., p. 814.
"	31–33.	C. " pusillus, Lx., p. 815.
"	34–35.	Rhabdocarpus late-costatus, Lx., p. 815.
"	36.	R. " inflatus, Lx., p. 815.
"	37–38.	R. " pachytesta, Lx., p. 816.
"	39–42.	R. " mamillatus, Lx., p. 815.
"	43–45, 62.	R. " subglobosus, Lx., p. 817.
"	46, 47.	R. " emarginatus, Lx., p. 818.
"	48, 49.	R. " abnormalis, Lx., p. 818.
"	50.	R. " apiculatus, Lx., p. 819.
"	51–57.	Trigonocarpus Adamsii, Lx., p. 820.
"	58–61.	T. " perpusillus, Lx., p. 820.
"	63–65.	T. " Schultzianus, G. & B., p. 819.
"	66–68.	Carpolithes minimus, St., p. 825.
"	69, 70.	C. " latior, Lx., p. 826.

PLATE CXI.

Figs.	1–3.	Trigonocarpus grandis, Lx., p. 821.
"	4–6.	T. " Dawesii, Ll. & H., p. 821.
"	7–13.	T. " Starkianus, Lx., p. 822.
"	14, 15.	T. multistriatus, Lx., p. 823.
"	16.	Cordaicarpus lineatus, Lx., p. 805.
"	17.	Carpolithes conicus, Lx., p. 824.
"	18.	C. " cerasiformis, St., p. 824.
"	19.	C. " Butlerianus, Lx., p. 824.
"	20, 21.	C. " granularis, St., p. 825.
"	22–24.	C. " perpusillus, Lx., p. 825.

Figs. 25, 26. Rhabdocarpus tenax, Lx., p. 818.
" 27–27b. Carpolithes transsectus, Lx., p. 826.
" 28. Sphenopteris plicata, Lx., p. 763.
" 29–29b. Pecopteris obsoleta, Lx., p. 758.
" 30–30b. P. " ornata, Lx., p. 760.

INDEX

OF

BOTANICAL NAMES

CONTAINED IN

REPORT P, VOLS. I, II, and III.

NOTE.—*When a * is prefixed to a page number it signifies that the* DESCRIPTION *is to be found* THERE. *On other pages where the name is found it is merely referred to, as synonyme, or otherwise.—References to figures on plates are in parenthesis.*

A.

B.

E.

G.

H.

L.

M.

R.

T

Page.

U.

V.

W.

X.

Z.

INDEX TO LOCALITIES MARKED IN THE LIST OF SPECIES.

A

B.

C.

D.

L.

M.

N.

O.

P.

R.

THE PUBLICATIONS

OF THE

SECOND GEOLOGICAL SURVEY OF PENNSYLVANIA.

REPORTS FOR 1874, 1875, 1876, 1877, 1878, 1879, 1880, 1881, 1882, AND 1883–1884.

Reports have been issued by the Board of Commissioners, and the prices thereof fixed in accordance with the law authorizing their publication, as follows:

ANTHRACITE COAL FIELDS.

A². SPECIAL REPORT TO THE LEGISLATURE UPON THE CAUSES, KINDS, AND AMOUNT OF WASTE IN MINING ANTHRACITE. By Franklin Platt, Assistant Geologist, with a chapter on the METHODS OF MINING. By John Price Wetherill, Mining Engineer. Illustrated by 35 figures of mining operations, a PLAN OF THE HAMMOND COAL BREAKER, on the Girard estate, and a SPECIMEN SHEET, scale 800 feet to 1 inch, $\frac{1}{9600}$ths of nature, illustrating the PROPOSED PLAN OF MAPPING THE ANTHRACITE FIELDS. By Chas. A. Ashburner, Assistant Geologist, 1881. 8 vo., pp. 134. Price, $1 10; postage, $0 12.

AC. REPORT ON THE MINING METHODS AND APPLIANCES used in the Anthracite Coal Fields. By H. M. Chance: with an atlas of 25 plates; 54 plates and 60 illustrations in the text. Price, $1 40; postage, $0 25.

AC. ATLAS. Coal Mining Plates I to XXV. By H. M. Chance. Price, $1 40; postage, $0 12.

AA. FIRST REPORT OF PROGRESS IN THE ANTHRACITE REGION, with a description of the Geology of the Panther Creek Basin, or Eastern End of the Southern Field. By Chas. A. Ashburner, Geologist in Charge; with an atlas of 13 sheets of maps and sections; 6 page plates, and 2 folded plates in the Report. Appendix A: Determination of the latitude and longitude of Wilkes Barre and Pottsville. By Prof. C. L. Doolittle. Appendix B: Theory of Stadia Measurements, with tables. By Arthur Winslow, assistant. 1883, 8 vo., pp. xlvii and 407. Price, $0 58; postage, $0 18.

AA. ATLAS SOUTHERN ANTHRACITE FIELD, VOLUME I, PANTHER CREEK* to accompany First Report of Progress AA, 1882. Contains 13 sheets, as follows: 3 mine sheets, 3 cross section sheets, 3 columnar section sheets, 1 topographical sheet, and 1 coal bed area sheet, all relating to the PANTHER CREEK BASIN IN CARBON AND SCHUYLKILL COUNTIES; also, 1 miscellaneous sheet, "General Preliminary Map, Anthracite Coal Fields," and 1 miscellaneous sheet containing chart, showing total annual production of Anthracite since 1820. Chas. A. Ashburner, Geologist in Charge, and A. W. Sheafer and Frank A. Hill, Assistant Geologists. Price, $1 50; postage, $0 12.

AA. ATLAS WESTERN MIDDLE ANTHRACITE FIELD, Part 1, 1884. Contains 11 sheets, as follows: 4 mine sheets between Delano and Locust Dale, 3 topographical sheets between Quakake Junction and Mount Carmel, and 4

NOTE.—*Single sheets of the Anthracite Survey, with the exception of those in the Panther Creek atlas, can be purchased by addressing Chas. A. Ashburner, Geologist in Charge, 907 Walnut street, Philadelphia.

cross-section sheets, all relating to the Mahanoy-Shamokin Basin in Schuylkill, Columbia, and Northumberland counties. In press. Chas. A. Ashburner, Geologist in Charge, and A. W. Sheafer and Bard Wells, Assistant Geologists. Price, $; postage, $.

AA. ATLAS NORTHERN FIELD, Part I, 1884. Contains 6 mine sheets between Wilkes Barre and Nanticoke, 3 cross-section sheets and — columnar section sheets, all relating to the Wyoming Basin in Luzerne county. In press Chas. A. Ashburner, Geologist in Charge, and Frank A. Hill, Assistant Geologist. Price, $; postage, $.

G². PART II. LOYALSOCK COAL BASIN, SULLIVAN COUNTY. By Franklin Platt. (See Reports Central Pennsylvania.)

BITUMINOUS COAL FIELDS AND SURROUNDING AREAS.

F. PART II. EAST BROAD TOP DISTRICT, HUNTINGDON COUNTY. By Chas. A. Ashburner. (See Reports Central Pennsylvania.)

G. REPORT OF PROGRESS IN BRADFORD AND TIOGA COUNTIES—1874-8. I. LIMITS OF THE CATSKILL AND CHEMUNG FORMATION. By Andrew Sherwood. II. Description of the BARCLAY, BLOSSBURG, FALL BROOK, ARNOT, ANTRIM, AND GAINES COAL FIELDS, and at the FORKS OF PINE CREEK IN POTTER COUNTY. By Franklin Platt. III. ON THE COKING OF BITUMINOUS COAL. By John Fulton. Illustrated with 2 colored *Geological* county *maps*, 3 page *plates*, and 35 *cuts*. 8 vo., pp. 271. Price, $1 00; postage, $0 12.

G². PART II. COAL BASINS, SULLIVAN AND LYCOMING COUNTIES. By Franklin Platt. (See Reports Central Pennsylvania.)

G³. REPORT OF PROGRESS IN 1876-9. The Geology of POTTER COUNTY, by Andrew Sherwood. Report on the COAL FIELDS, by Franklin Platt, with a colored geological map of the county, two folded plates, and two page plates of sections. 8 vo., pp. 120. Price, $0 58; postage, $0 08.

G⁴. REPORT OF PROGRESS. Part I. GEOLOGY OF CLINTON COUNTY. Part II. A special study of the CARBONIFEROUS and DEVONIAN STRATA along the West Branch of Susquehanna River. By H. Martyn Chance. Included in this report is a description of the RENOVO COAL BASIN, by Chas. A. Ashburner, and notes on the TANGASCOOTACK COAL BASIN in Centre and Clinton Counties, by Franklin Platt. Price, $1 05; postage, $0 12.

H. REPORT OF PROGRESS IN THE CLEARFIELD AND JEFFERSON DISTRICT OF THE BITUMINOUS COAL FIELDS of Western Pennsylvania—1874. By Franklin Platt. 8 vo., pp. 296, illustrated by 139 *cuts*, 8 *maps*, and 2 *sections*. Price in paper, $1 50; postage, $0 13.

H². REPORT OF PROGRESS IN THE CAMBRIA AND SOMERSET DISTRICT OF THE BITUMINOUS COAL FIELDS of Western Pennsylvania—1875. By F. and W. G. Platt. Pp. 194, illustrated with 84 *wood-cuts*, and 4 *maps* and *sections*. Part I. Cambria. Price, $1 00; postage, $0 12.

H³. REPORT OF PROGRESS IN THE CAMBRIA AND SOMERSET DISTRICT OF THE BITUMINOUS COAL FIELDS of Western Pennsylvania—1876. By F. and W. G. Platt. Pp. 348, illustrated by 110 *wood-cuts* and 6 *maps* and *sections*. Part II. Somerset. Price, $0 85; postage, $0 18.

H⁴. REPORT OF PROGRESS IN INDIANA COUNTY—1877. By W. G. Platt. Pp. 316. With a colored map of the county. Price, $0 80; postage, $0 14.

H⁵. REPORT OF PROGRESS IN ARMSTRONG COUNTY—1879. By W. G. Platt. Pp. 338. With a colored map of the county. Price, $0 75; postage, $0 16.

H[6]**.** REPORT OF PROGRESS IN JEFFERSON COUNTY—1880; with colored map of county. By W. G. Platt. Price, $0 60; postage, $0 12.

H.[7] A REVISION OF THE BITUMINOUS COAL MEASURES OF CLEARFIELD COUNTY—1884; with a colored geological county map; outcrop map of the Houtzdale Basin, and coal bed sections in the text. By H. M. Chance. Price, $; postage, $.

I[4]. QUAKER HILL COAL BASIN, WARREN COUNTY. By John F. Carll. (Sea Reports Petroleum Fields.)

K. REPORT ON GREENE AND WASHINGTON COUNTIES—1875, Bituminous Coal Fields. By J. J. Stevenson, 8 vo., pp. 420, illustrated by 3 *sections* and 2 county *maps*, showing the depth of the Pittsburgh and Waynesburg coal bed beneath the surface at numerous points. Price in paper, $0 65; postage, $0 16.

K[2]**.** REPORT OF PROGRESS IN THE FAYETTE AND WESTMORELAND DISTRICT OF THE BITUMINOUS COAL FIELDS of Western Pennsylvania—1876. By J. J. Stevenson; pp. 437, illustrated by 50 *wood-cuts* and 3 county *maps*, colored. Part I. Eastern Allegheny County, and Fayette and Westmoreland Counties, west from Chestnut Ridge. Price, $1 40; postage, $0 20.

K[3]**.** REPORT OF PROGRESS IN THE FAYETTE AND WESTMORELAND DISTRICT OF THE BITUMINOUS COAL FIELDS of Western Pennsylvania—1877. By J. J. Stevenson. Pp. 331. Part II. The LIGONIER VALLEY. Illustrated with 107 *wood-cuts*, 2 *plates*, and 2 county *maps*, colored. Price, $1 40; postage, $0 16.

M, M[2] and **M**[3]**.** REPORTS OF PROGRESS IN THE LABORATORY. By Andrew S. McCreath. Contains coal analyses.

P. REPORT AND ATLAS OF THE COAL FLORA. By Leo Lesquereux.

P[2]. REPORT OF THE PERMIAN AND UPPER CARBONIFEROUS FLORA. By Wm. M. Fontaine and I. C. White. (See Miscellaneous Reports.)

Q. REPORT OF PROGRESS IN THE BEAVER RIVER DISTRICT OF THE BITUMINOUS COAL FIELDS OF WESTERN PENNSYLVANIA. By I. C. White. Pp. 337, illustrated with 3 *Geological maps* of parts of Beaver, Butler, and Allegheny Counties, and 21 *plates of vertical sections.* 1875. • Price, $1 40; postage, $0 20.

Q[2]**.** REPORT OF PROGRESS IN 1877. The Geology of LAWRENCE COUNTY, to which is appended a Special Report on the CORRELATION OF THE COAL MEASURES in Western Pennsylvania and Eastern Ohio. 8 vo., pp. 336, with a *colored Geological Map* of the county, and 134 vertical *sections.* By I. C. White. Price, $0 70; postage, $0 15.

Q[3]**.** REPORT OF PROGRESS IN 1878. The Geology of MERCER COUNTY, by I. C. White, with a colored geological map of county, and 119 vertical sections. 8 vo., pp. 233. Price, $0 60; postage, $0 11.

R. REPORT OF PROGRESS. The Geology of MCKEAN COUNTY, and its connection with that of CAMERON, ELK, and FOREST, with Atlas containing 8 sheets of maps and sections. By Chas. A. Ashburner. Price, $1 70; postage, $0 22.

T. COAL MEASURES, BLAIR COUNTY. By Franklin Platt.

T[2]**.** COAL MEASURES, BEDFORD AND FULTON COUNTIES. By J. J. Stevenson. (See Reports Central Pennsylvania.)

V. REPORT OF PROGRESS—1878. Part I. The Northern Townships of Butler county. Part II. A special survey made in 1875, along the Beaver and Shenango rivers, in BEAVER, LAWRENCE, and MERCER COUNTIES. 8 vo., pp. 248, with 4 *maps*, 1 profile *section* and 154 vertical *sections.* By H. Martyn Chance. Price, $0 70; postage, $0 15.

V^2. REPORT OF PROGRESS IN 1879. 8 vo., pp. 232. The Geology of CLARION COUNTY, by H. Martyn Chance, with colored geological map of county, a map of the Anticlinals and OIL BELT, a contoured map of the Old River Channel at Parker, 83 local sections figured in the text, and 4 page plates. Price, $0 43: postage, $0 12.

PETROLEUM FIELDS.

I. REPORT OF PROGRESS IN THE VENANGO COUNTY DISTRICT—1874. By John F. Carll. With observations on the Geology around Warren, by F. A. Randall; and Notes on the Comparative Geology of North-eastern Ohio and North-western Pennsylvania, and Western New York, by J. P. Lesley. 8 vo., pp. 127, with 2 *maps*, a long *section*, and 7 *cuts* in the text. Price in paper, $0 60; postage, $0 05.

I^2. REPORT OF PROGRESS, OIL WELLS, RECORDS, AND LEVELS—1876-7. By John F. Carll. Pp. 398. Published in advance of Report of Progress, III. Price, $0 60; postage, $0 18.

I^3. REPORT OF PROGRESS—1875 to 1879. Geology of the OIL REGIONS OF WARREN, VENANGO, CLARION, AND BUTLER COUNTIES, including surveys of the GARLAND and PANAMA CONGLOMERATES in Warren and Crawford counties, and in Chautauqua county, New York, with descriptions of oil well rig and tools, and a discussion of the preglacial and postglacial drainage of the LAKE ERIE COUNTRY; with Atlas. With maps and charts of Oil Regions. By John F. Carll. Price, $2 30; postage, $0 30.

I^4. GEOLOGICAL REPORT OF WARREN COUNTY AND NEIGHBORING OIL REGIONS, with additional oil well records—1880-3. By John F. Carll, with colored geological map of Warren county, two sheets of oil well sections, and a map of the Warren oil region. 439 pages. Price, $1 12; postage, $0 20.

J. SPECIAL REPORT ON THE PETROLEUM OF PENNSYLVANIA—1874, its Production, Transportation, Manufacture, and Statistics. By Henry E. Wrigley. To which are added a Map and Profile of a line of levels through Butler, Armstrong, and Clarion Counties, by D. Jones Lucas: and also a Map and Profile of a line of levels along Slippery Rock Creek, by J. P. Lesley. 8 vo., pp. 122; 5 *maps* and *sections*, a *plate* and 5 *cuts*. Price in paper, $0 75: postage, $0 06.

K. DUNKARD CREEK OIL DISTRICT, GREENE COUNTY. By J. J. Stevenson. (See Reports Bituminous Coal Fields.)

L. Appendix II. A REPORT ON THE USE OF NATURAL GAS IN IRON MANUFACTURE. By John B. Pearse. (See Miscellaneous Reports.)

Q^2. DESCRIPTION OF OIL MEASURES IN AND ADJACENT TO LAWRENCE COUNTY. By I. C. White. (See Reports Bituminous Coal Fields.)

Q^4. DESCRIPTION OF OIL MEASURES IN AND ADJACENT TO EIRE AND CRAWFORD COUNTIES. By I. C. White. (See Reports North-western Pennsylvania.)

R. DESCRIPTION OF THE BRADFORD OIL DISTRICT IN MCKEAN COUNTY, with a reference to the probable position of the Oil Sands in Elk county. By Chas. A. Ashburner. (See Reports Bituminous Coal Fields.)

V^2. DESCRIPTION OF THE OIL MEASURES IN CLARION COUNTY. By H. M. Chance. (See Reports Bituminous Coal Fields.)

NORTH-WESTERN PENNSYLVANIA.

Q^4. REPORT OF PROGRESS—1879. The Geology of ERIE AND CRAWFORD COUNTIES, with tables of barometric heights in each township, and notes on the place of the SHARON CONGLOMERATE in the Palæozoic series. By I. C.

White. Also, the discovery of the PREGLACIAL OUTLET OF LAKE ERIE, with two maps of the Lake Region. By J. W. Spencer, Ph. D. Price, $1 17; postage, $0 18.

I, I^{2}, I^{3}, I^{4}, Q^{3}, V, V^{2} and **R.** PETROLEUM REGION REPORTS. By John F. Carll, I. C. White, H. M. Chance, and Chas. A. Ashburner.

CENTRAL PENNSYLVANIA.

F. REPORT OF PROGRESS IN THE JUNIATA DISTRICT on Fossil Iron Ore Beds of Middle Pennsylvania. By John H. Dewees. With a report of the AUGHWICK VALLEY AND EAST BROAD TOP DISTRICT. By C. A. Ashburner. 1874-8. Illustrated with 7 *Geological maps* and 19 *sections*. 8 vo., pp. 305. Price, $2 55; postage, $0 20.

G. REPORT OF PROGRESS IN BRADFORD AND TIOGA COUNTIES. By Andrew Sherwood. (See Reports Bituminous Coal Fields.)

G^{2}. REPORT OF PROGRESS. GEOLOGY OF LYCOMING AND SULLIVAN COUNTIES. I. Field Notes by Andrew Sherwood. II. Coal Basins, by Franklin Platt. With two colored geological county maps and numerous illustrations. 8 vo., pp. 268. Price, $1 06; postage, $0 14.

G^{1} REPORT OF PROGRESS IN CLINTON COUNTY. By H. M. Chance. (See Reports Bituminous Coal Fields.)

G^{7}. REPORT OF PROGRESS. THE GEOLOGY IN THE SUSQUEHANNA RIVER REGION IN THE SIX COUNTIES OF WYOMING, LACKAWANNA, LUZERNE, COLUMBIA, MONTOUR, AND NORTHUMBERLAND. By I. C. White. With a colored Geological Map in 2 sheets; and 31 page plates in text. Pp. 464. Price, $0 85; postage, $0 20.

T. REPORT OF PROGRESS. Geology of BLAIR COUNTY, with 35 illustrations and an Atlas of 14 sheets of the colored map of MORRISON'S COVE, &c.; 1 index sheet, and 2 sheets of colored sections. By Franklin Platt. Price of Report and Atlas, $4 55; postage, $0 28.

T^{2}. REPORT OF PROGRESS—1882. The geology of BEDFORD AND FULTON COUNTIES. By J. J. Stevenson. 8 vo., pp. 382. Illustrated with 2 colored geological *maps*. Price, $0 80; postage, $0 20.

NORTH-EASTERN PENNSYLVANIA.

G^{5}. REPORT OF PROGRESS. THE GEOLOGY OF SUSQUEHANNA COUNTY AND WAYNE COUNTY. By I. C. White. Pp. 243, with Geological map and 58 sections. Price, $0 70; postage, $0 12.

G . REPORT OF PROGRESS, 1881. THE GEOLOGY OF PIKE AND MONROE COUNTIES. By I. C. White. 8 vo., pp. 407. Illustrated with colored *Geological* county *maps*, a *map* of glacial scratches, and 7 small *sections*. Also special surveys of the DELAWARE AND LEHIGH WATER GAPS. By H. M. Chance, with 2 *contoured maps* of Water Gaps, and 5 *detailed sections*. Price, $1 15; postage, $0 15.

G^{7}. THE GEOLOGY IN THE SUSQUEHANNA (NORTH BRANCH) RIVER REGION IN THE SIX COUNTIES OF WYOMING, LACKAWANNA, LUZERNE, COLUMBIA, MONTOUR, NORTHUMBERLAND, (exclusive of ANTHRACITE REGION.) By I. C. White. (See Reports Central Pennsylvania.)

G. REPORT OF PROGRESS IN BRADFORD AND TIOGA COUNTIES. By Andrew Sherwood. (See Reports Bituminous Coal Fields.)

A^{2}, AA, and **AC.** ANTHRACITE REGION REPORTS. By Franklin Platt, Chas. A. Ashburner, and H. M. Chance. (See Reports Anthracite Coal Fields.)

SOUTH-EASTERN PENNSYLVANIA.

C. REPORT OF PROGRESS ON YORK AND ADAMS COUNTIES—1874. By Persifor Frazer. 8vo., pp. 198, illustrated by 8 *maps* and *sections* and other illustrations. Price in paper, $0 85; postage, $0 10.

C^2. REPORT OF PROGRESS IN THE COUNTIES OF YORK, ADAMS, CUMBERLAND, AND FRANKLIN—1875. Illustrated by *maps* and *cross-sections*, showing the Magnetic and Micaceous Ore Belt near the western edge of the Mesozoic Sandstone and the two Azoic systems constituting the mass of the South Mountains, with a preliminary discussion on the DILLSBURG ORE BED and catalogue of specimens collected in 1875. By Persifor Frazer. Price, $1 25; postage, $0 12.

C^3. REPORT OF PROGRESS IN 1877. The Geology of LANCASTER COUNTY, with an atlas containing a colored geological map of the county, local map of the GAP NICKEL MINE, map and sections of the East Bank of Susquehanna River; other geological sections across the county, and geological colored maps of York and Lancaster counties. By Persifor Frazer. 8 vo., pp. 350. Price of Report and Atlas, $2 20; postage, $0 25.

C^4. GEOLOGY OF CHESTER COUNTY, after the surveys of Henry D. Rogers, Persifor Frazer and Charles E. Hall, edited by J. P. Lesley—with a colored geological map of the county, three lithographic plates and maps, and sections in the text. Price, $0 75; postage, $0 18.

C^6. REPORT OF PROGRESS. GEOLOGY OF PHILADELPHIA COUNTY, AND OF THE SOUTHERN PARTS OF MONTGOMERY AND BUCKS. By Charles E. Hall. Pp. 145, with Geological map, sheet of colored cross-sections, and 24 page cuts. Price, $1 65; postage, $0 13.

D. REPORT OF PROGRESS IN THE BROWN HEMATITE ORE RANGES OF LEHIGH COUNTY—1874, with descriptions of mines lying between Emaus, Alburtis, and Fogelsville. By Frederick Prime, Jr. 8vo., pp. 73, with a contour-line *map* and 8 *cuts*. Price in paper, $0 50; postage, $0 04.

D^2. THE BROWN HEMATITE DEPOSITS OF THE SILURO-CAMBRIAN LIMESTONES OF LEHIGH COUNTY, lying between Shimersville, Millerstown, Schenecksville, Ballietsville, and the Lehigh river—1875-6. By Frederick Prime, Jr. 8 vo., pp. 99, with 5 *map-sheets* and 5 *plates*. Price, $1 60; postage, $0 12.

D^3. VOL. I. REPORT OF PROGRESS. GEOLOGY OF LEHIGH AND NORTHAMPTON COUNTIES. General introduction, by J. P. Lesley. Slate Belt and Quarries, by R. A. Sanders. Water Gaps, by H. M. Chance. Limestone Belt and Iron Ore Mines, by F. Prime. South Mountain Rocks, by F. Prime. Itinerary Survey, by C. E. Hall. Three lithograph and 3 artotype views of quarries, and an atlas. Pp. 283. Price, $0 65; postage, $0 13.

D^3. VOL. II. PART I. REPORT OF PROGRESS. GEOLOGY OF THE SOUTH MOUNTAIN BELT OF BERKS COUNTY. By E. V. D'Invilliers. Illustrated by 18 page plates in the text, and by the maps in the Atlas. Pp. 441. Price, $0 55; postage, $0 18.

D^3. Volumes I and II, ATLAS, containing a colored contour map of Southern Northampton on 6 sheets, a contour map of the mountain on 18 sheets, a geological index map on 1 sheet, a colored geological map of NORTHAMPTON AND LEHIGH COUNTIES, and 4 maps of IRON MINES IN BERKS COUNTY. Price, $2 80; postage, $0 17.

D^5. MAPS OF ADAMS, FRANKLIN, AND CUMBERLAND COUNTIES. South Mountain sheets A^1, A^2, B^1 and B^2. By A. E. Lehman. Price, $1 25; postage, $0 08.

E. Special Report on the Trap Dykes and Azoic Rocks of South-Eastern Pennsylvania—1875. Part I, Historical Introduction. By T. Sterry Hunt. 8 vo., pp. 253. Price, $0 48; postage, $0 12.

MISCELLANEOUS REPORTS.

A. Historical Sketch of Geological Explorations in Pennsylvania and other States. By J. P. Lesley. With appendix, containing Annual Reports for 1874 and 1875; pp. 226, 8vo. Price in paper, $0 25; postage, $0 06.

B. Preliminary Report of the Mineralogy of Pennsylvania—1874. By Dr. F. A. Genth. With appendix on the hydro-carbon compounds, by Samuel P. Sadtler. 8vo., pp. 206, with *map* of the State for reference to counties. Price in paper, $0 50; postage, $0 08. Price in cloth, $0 75; postage, $0 10.

L. 1875—Special Report on the Coke Manufacture of the Youghiogheny River Valley in Fayette and Westmoreland Counties, with Geological Notes of the Coal and Iron Ore Beds, from Surveys, by Charles A. Young; by Franklin Platt. To which are appended: I. A Report on Methods of Coking, by John Fulton. II. A Report on the use of Natural Gas in the Iron Manufacture, by John B. Pearse, Franklin Platt, and Professor Sadtler. Pp. 252. Price, $1 00; postage, $0 13.

M. Report of Progress in the Laboratory of the Survey at Harrisburg—1874-5. By Andrew S. McCreath. 8 vo., pp. 105. Price in paper, $0 50: postage, $0 05.

M². Second Report of Progress in the Laboratory of the Survey, at Harrisburg, by Andrew S. McCreath—1876-8, including I. Classification of Coals, by Persifor Frazer. II. Firebrick Tests, by Franklin Platt. III. Notes on Dolomitic Limestones, by J. P. Lesley. IV. Utilization of Anthracite Slack, by Franklin Platt. V. Determination of Carbon in Iron or Steel, by A. S. McCreath. With 3 indexes, plate, and 4 page plates. Pp. 438. Price in cloth, $0 65; postage, $0 18.

M³. Third Report of Progress in the Laboratory of the Survey, at Harrisburg. Analyses, &c., &c. By Andrew S. McCreath. Pp. 126, with 2 indexes and map. Price, $0 40; postage, $0 10.

N. Report of Progress—1875-6-7. Two Hundred Tables of Elevation above Tide-Level of the Railroad Stations, Summits and Tunnels; Canal Locks and Dams, River Riffles, &c., in and around Pennsylvania; with *map*; pp. 279. By Charles Allen. Price, $0 70; postage, $0 15.

O. Catalogue of the Geological Musuem—1874-5-6-7. By Charles E. Hall. Part I. Collection of Rock Specimens. Nos. 1 to 4,264. Pp. 217. Price, $0 40; postage, $0 10.

O². Catalogue of the Geological Museum. By Charles E. Hall. Part II. 1. Collections of rock specimens, Nos. 4265 to 8974. 2. Palæontological specimens. Price, $0 40; postage, $0 12.

P. 1879—Report and Atlas of the Coal Flora of Pennsylvania and of the Carboniferous Formation throughout the United States. By Leo Lesquereux. Price of Report, $0 80; postage, $0 28. Price of Atlas, $3 35; postage, $0 22.

P². The Permian or Upper Carboniferous Flora of West Virginia and S. W, Pennsylvania, with 38 plates. By Wm. M. Fontaine, M. A., and I. C. White, A. M. Price, $2 25; postage, $0 17.

Other Reports of the Survey are in the hands of the State Printer, and will soon be published.

The sale of the reports is conducted in accordance with the provisions of Section 10 of the Act of the 14th day of May, 1874, which directs that copies of the Reports, with all maps and supplements, shall be furnished *at cost of publication to all applicants for them.*

All the printed volumes and maps in stock have been transferred by the Board of Commissioners to the Department of Internal Affairs, where the sales thereof will hereafter be conducted.

Communications relating to the work of the Survey should be addressed to J. P. Lesley, State Geologist, No. 1008 Clinton street, Philadelphia, and those intended for the Board of Commissioners, to William A. Ingham, Secretary, No. 907 Walnut street, Philadelphia.

All letters and orders concerning the purchase of Reports and remittances for the same, should be addressed to,

J. SIMPSON AFRICA,
Secretary of Internal Affairs,
Harrisburg. Pa.

April 1, 1884.

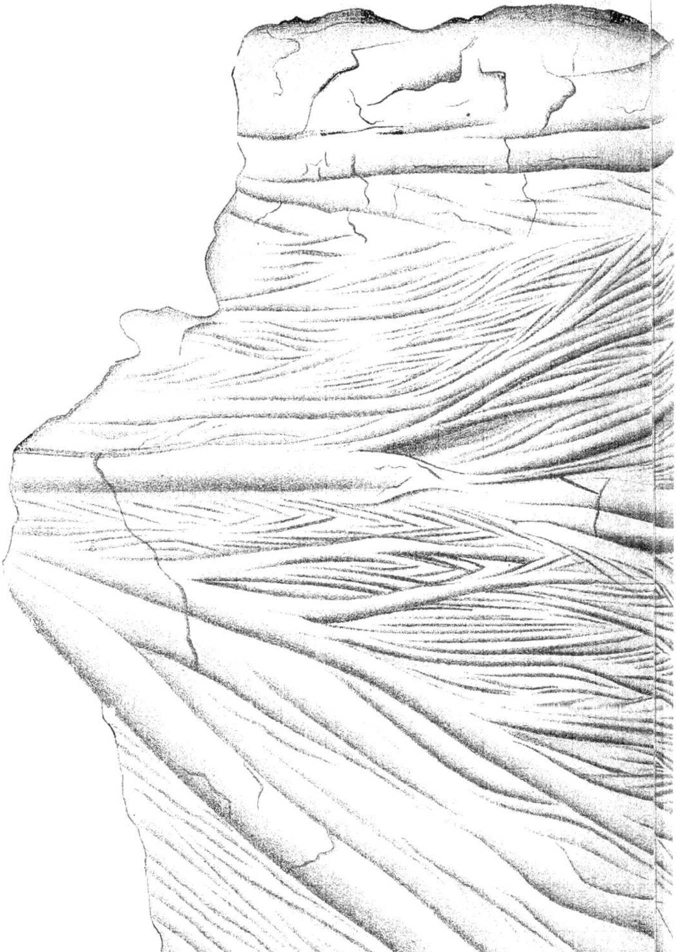

1.

Fred. Earhart, del.

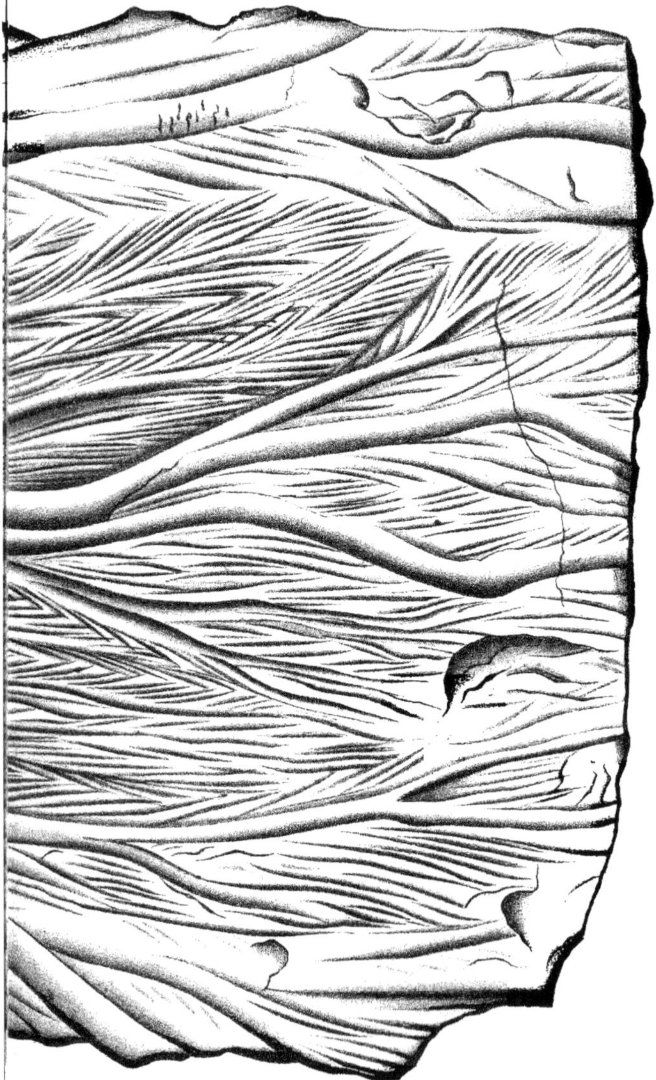

Leon S. Hart, Lith. Harrisburg, Pa.

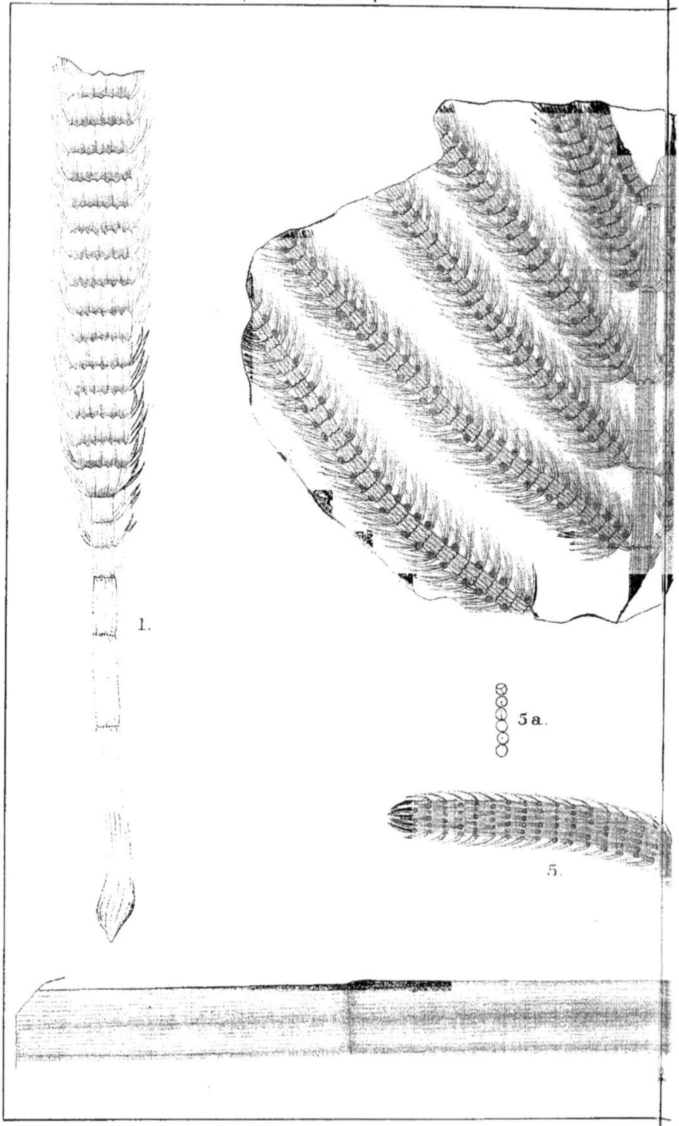

Fred. Earhart, del.

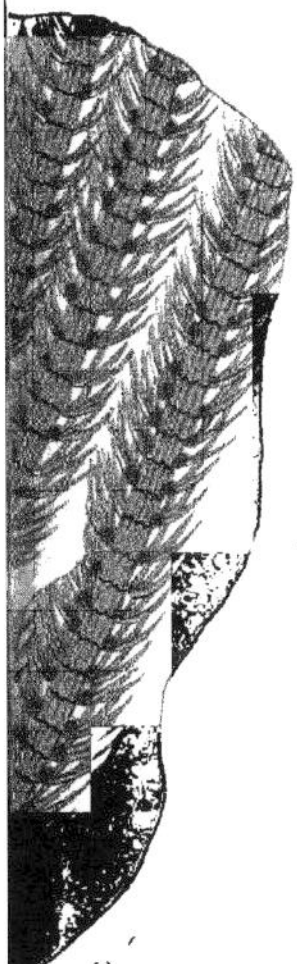

3.

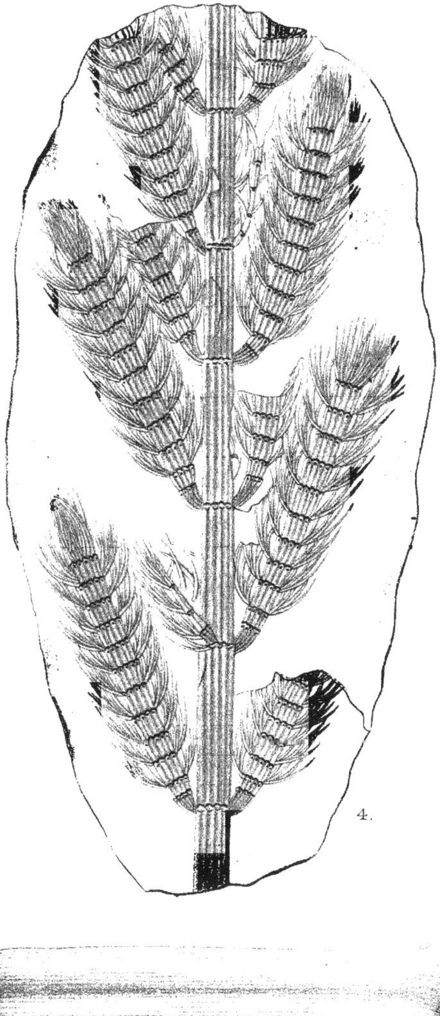

4.

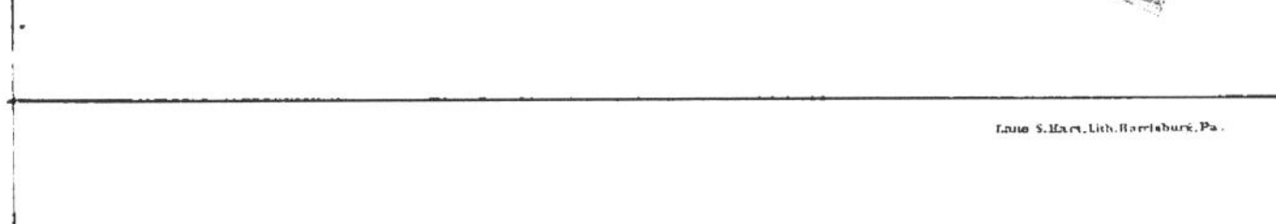

Lane S. Hart, Lith. Harrisburg, Pa.

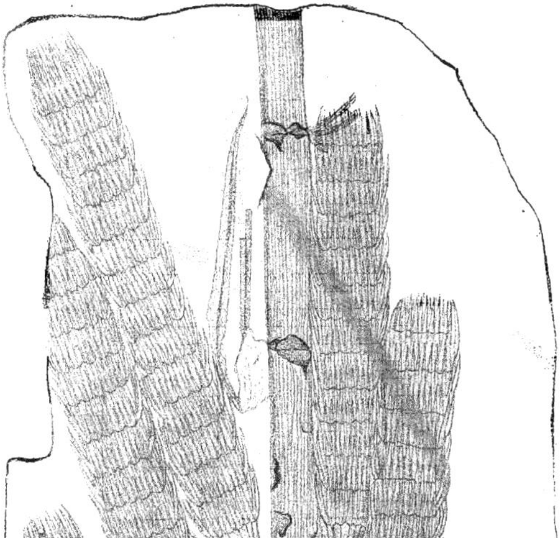

1.

Fred. Earhart, del.

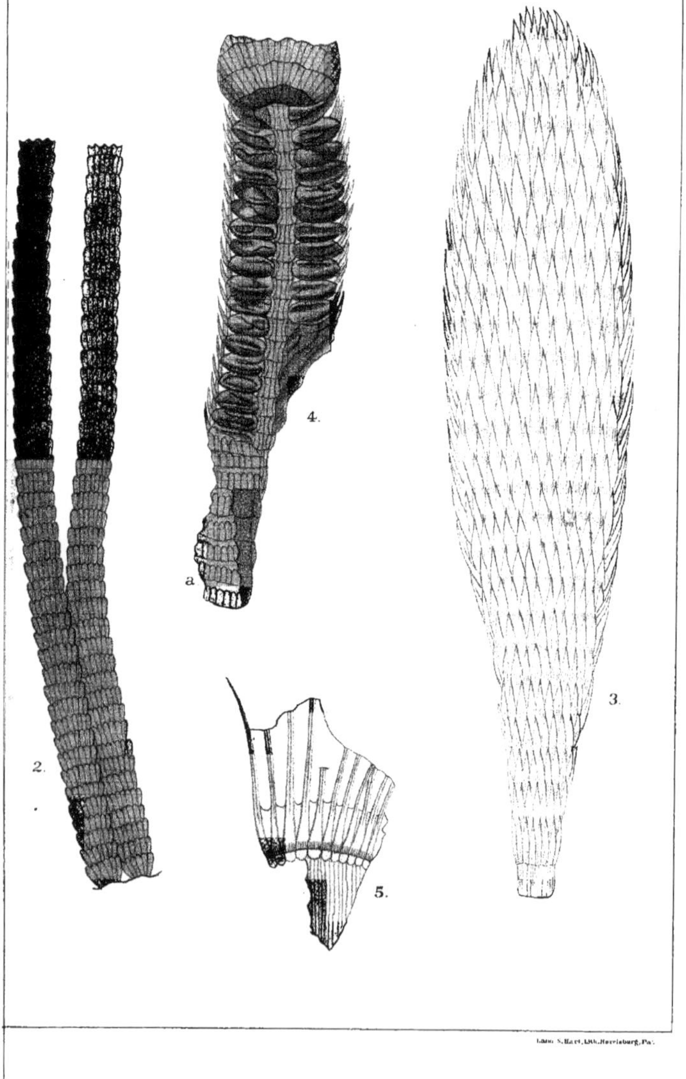

[illegible] S. Hart, Lith. Harrisburg, Pa.

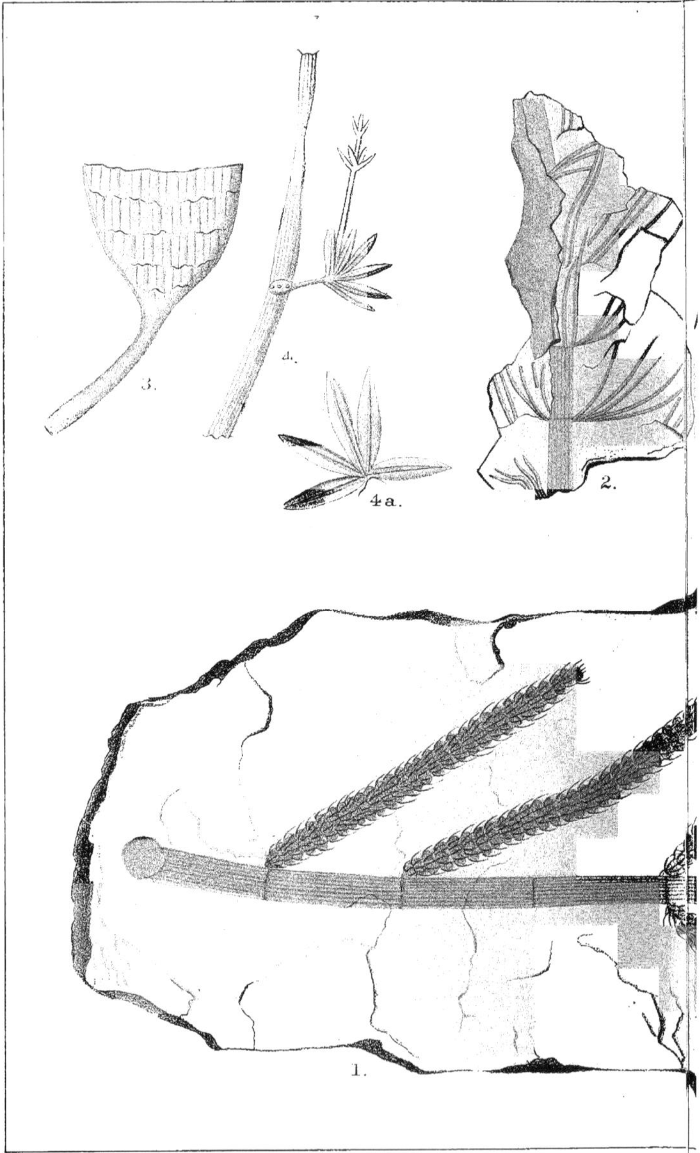

Fred. Earhart, del.

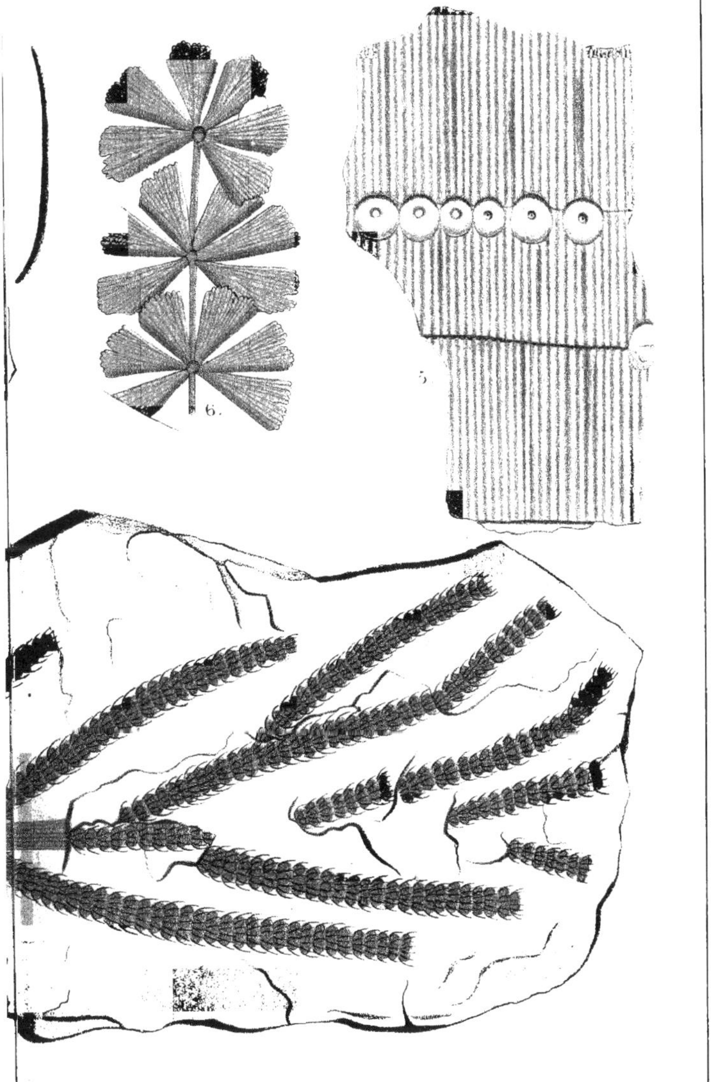

Lane S. Hart, Lith. Harrisburg, Pa

Second Geol. Sur. Pennsylvania, L. Lesquereux 1883.

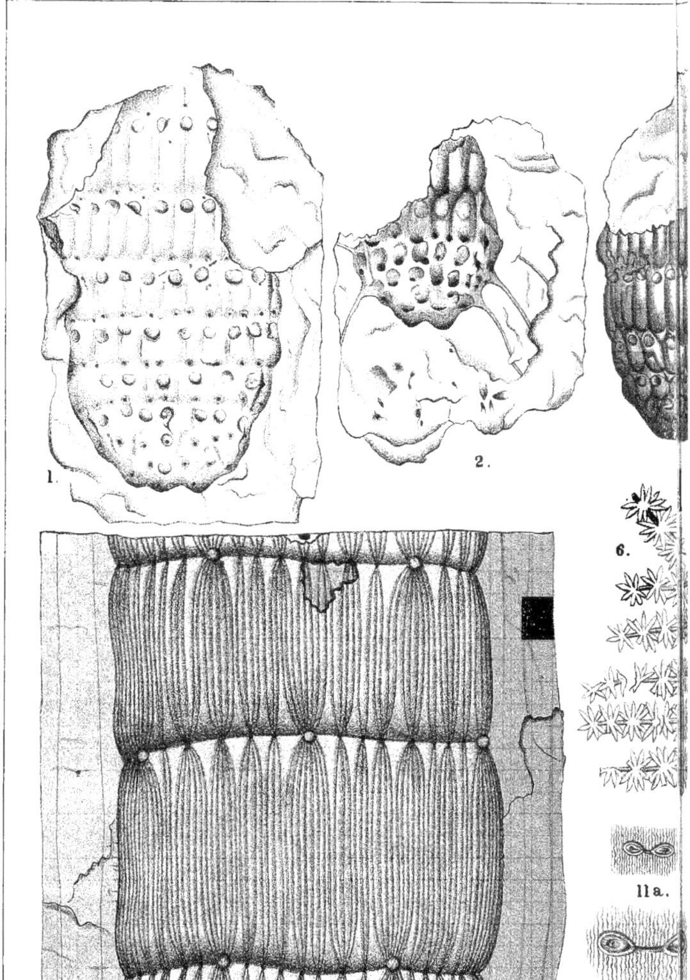

Fred. Earhart, del.

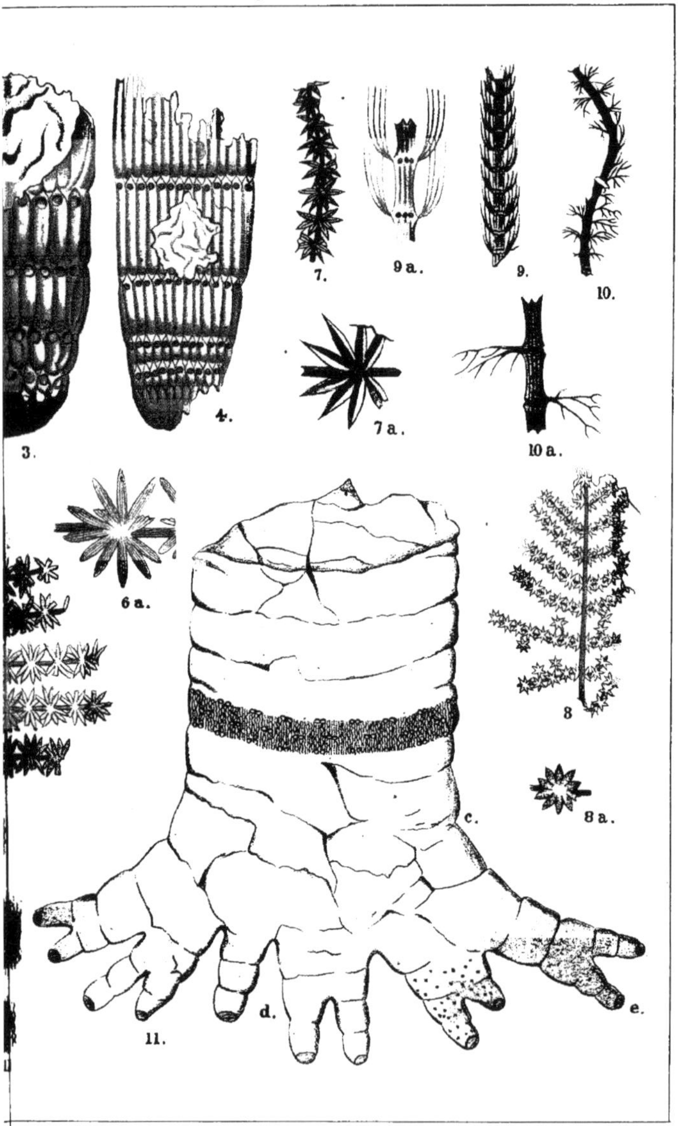

Lane S. Hart, Lith. Harrisburg, Pa.

Second Geol. Sur. Pennsylvania. L. Lesquereux 1883.

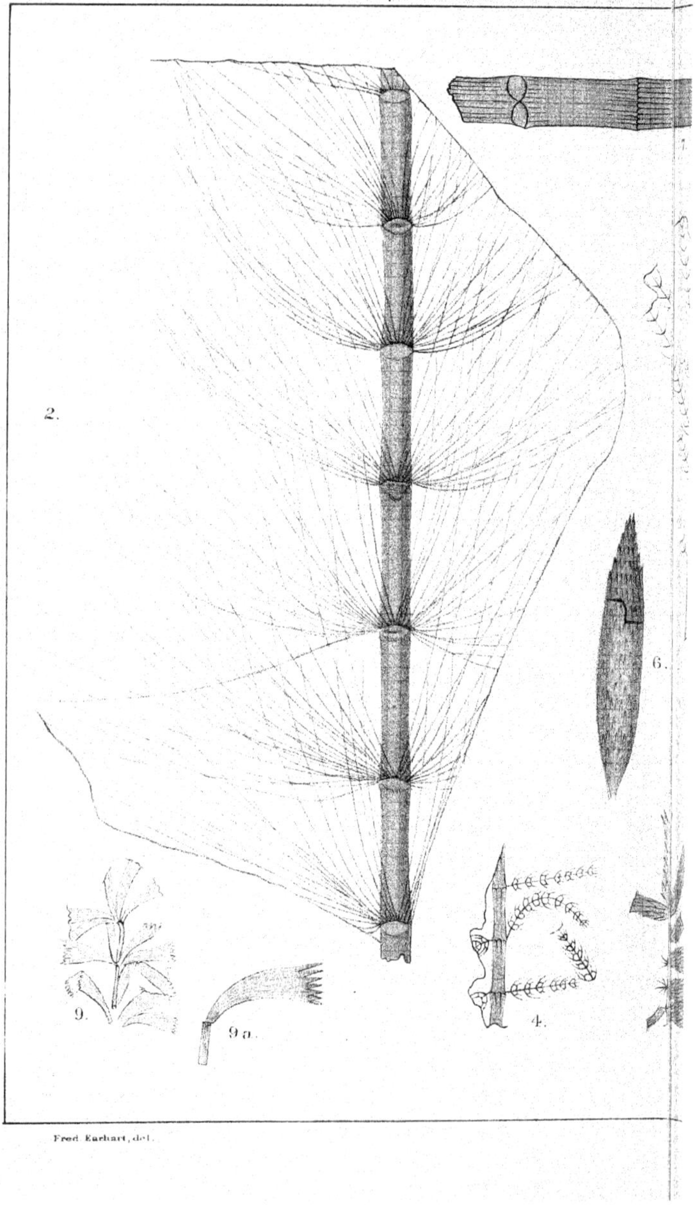

Fred. Earhart, del.

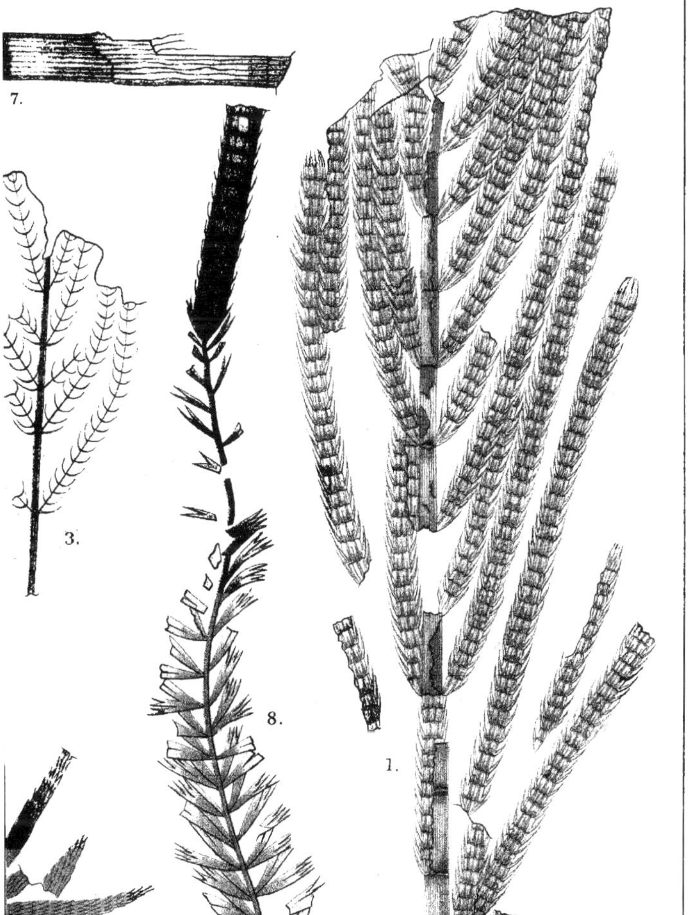

Lane S. Hart, Lith. Harrisburg, Pa.

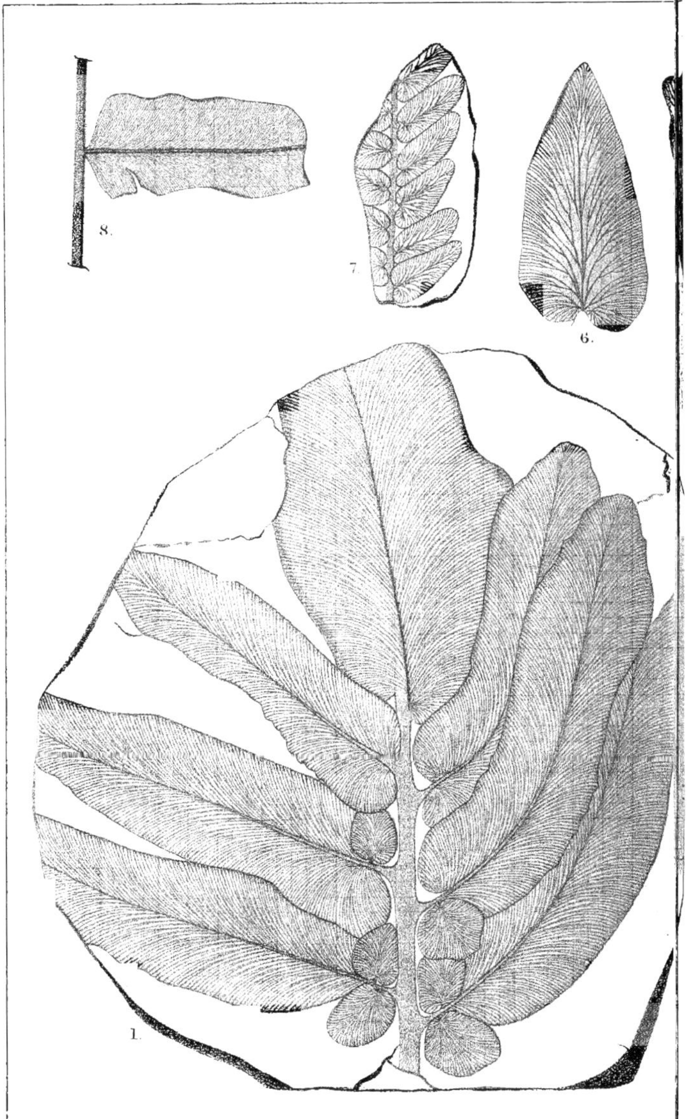

Fred. Earhart, del.

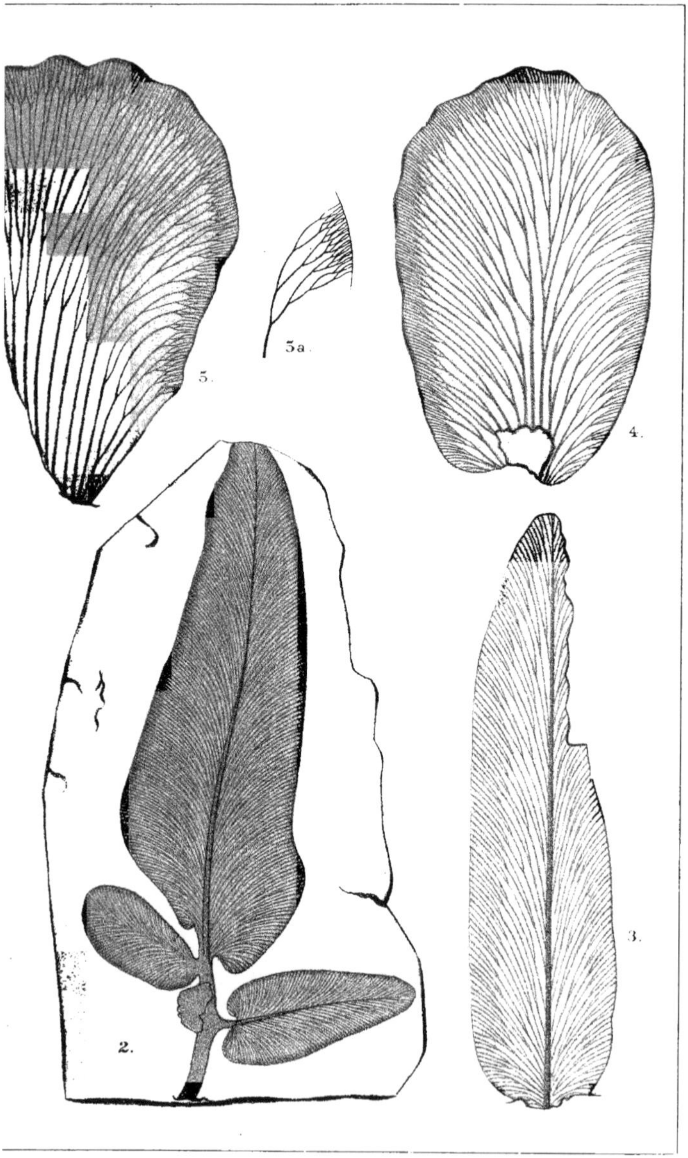

Lane S. Hart, Lith. Harrisburg, Pa.

Second Geol. Sur. Pennsylvania, L. Lesquereux 1883.

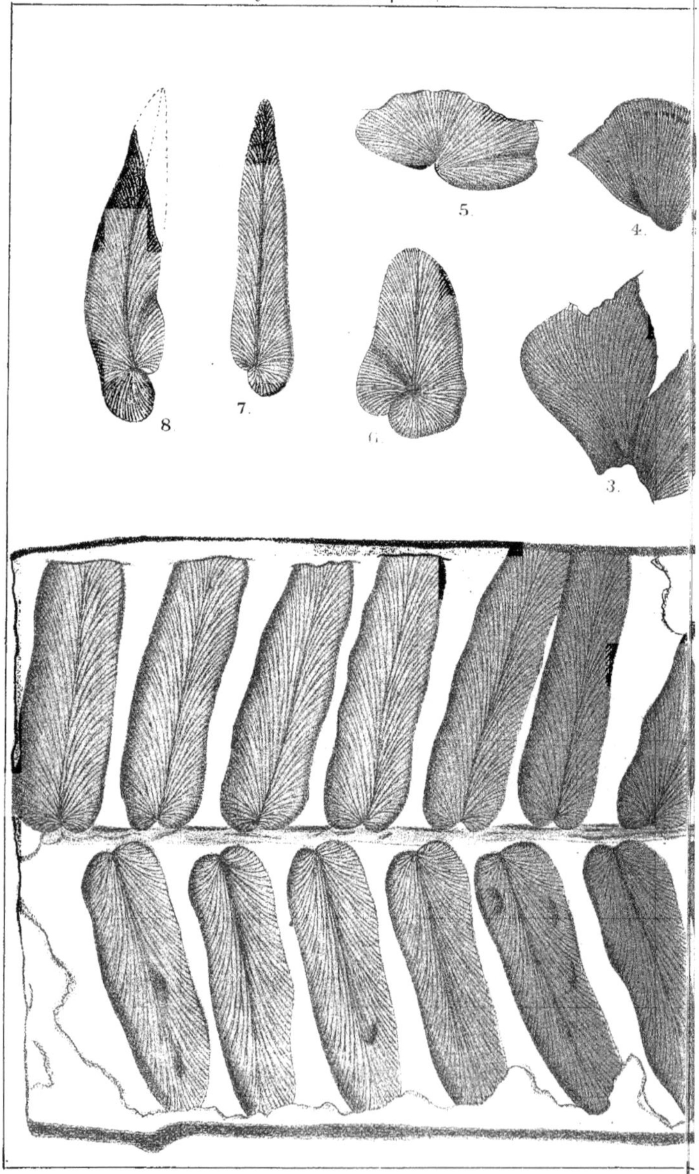

Fred. Earhart, del.

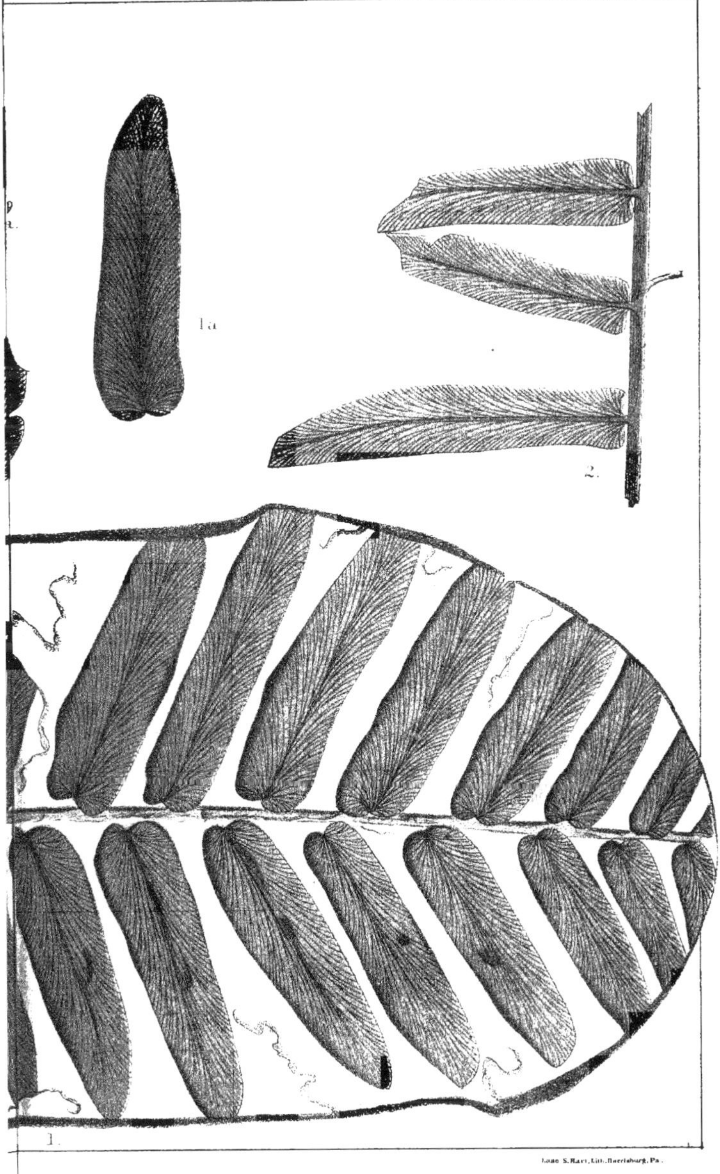
1 a
2.
1.

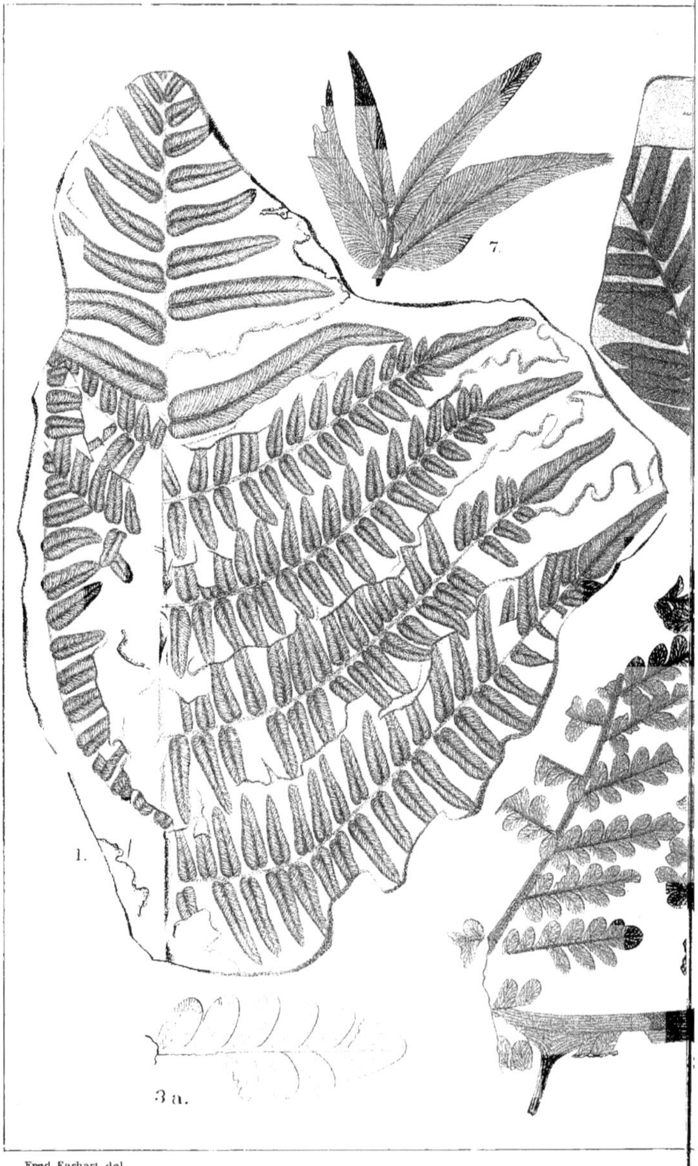

Fred. Eachart, del.

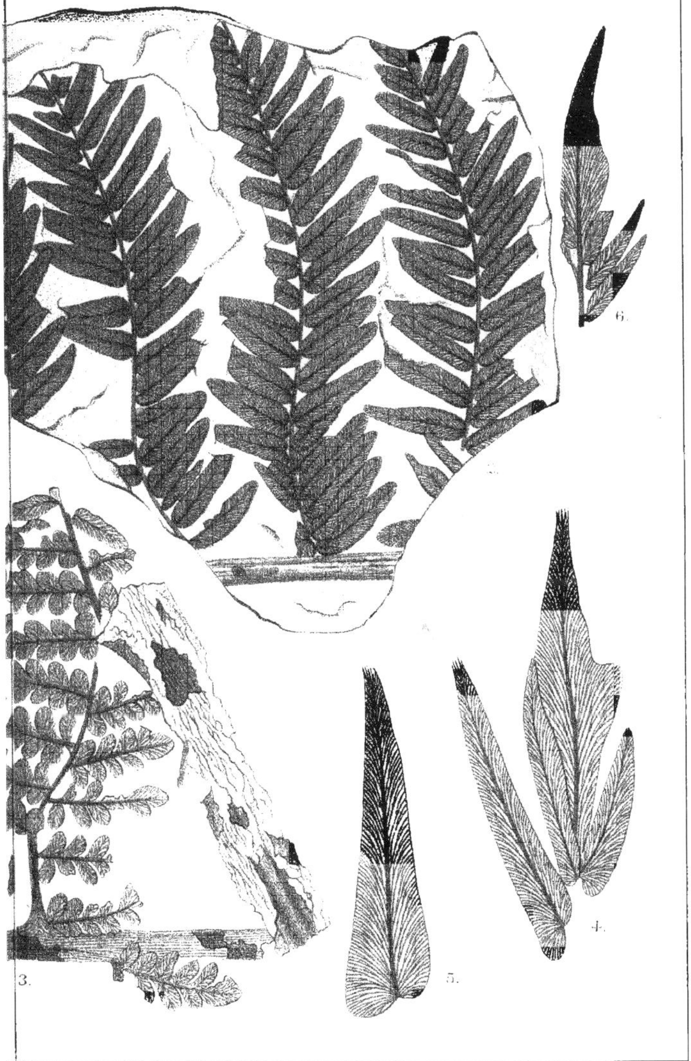

Lane S. Hart, Lith. Harrisburg, Pa.

Second Geol. Sur. Pennsylvania, L. Lesquereux 1883.

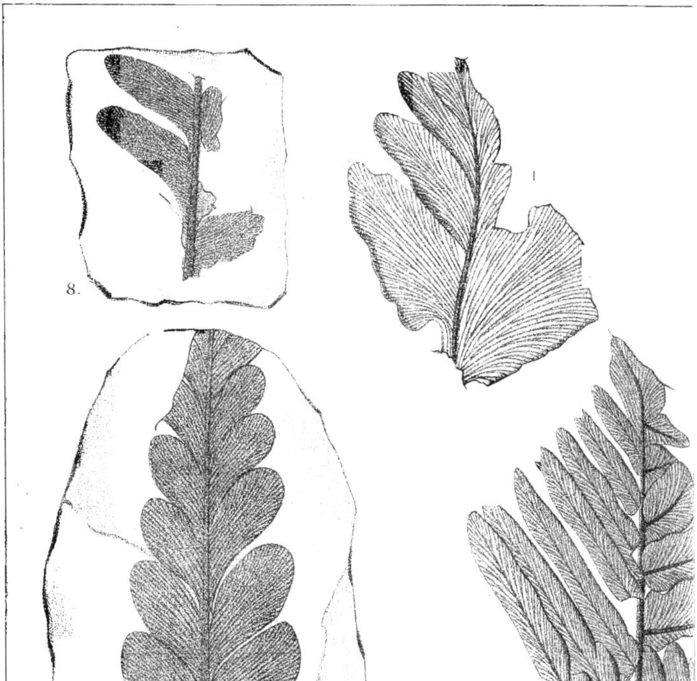

Fred. Earhart, del.

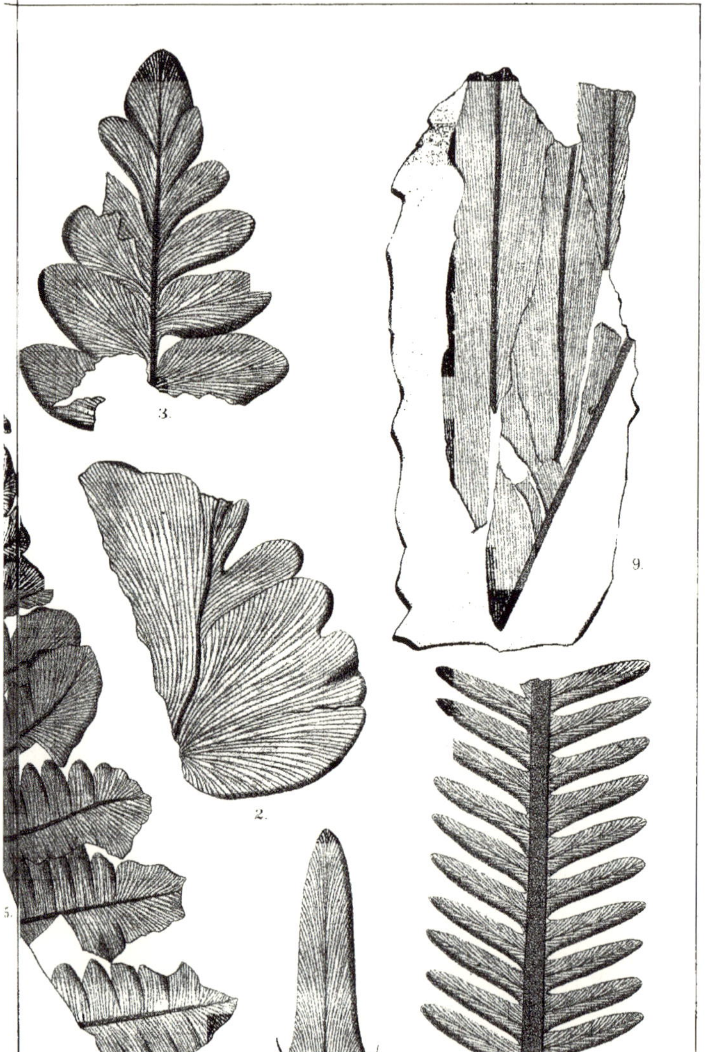

Lith. S. Hart, Lith. Harrisburg, Pa.

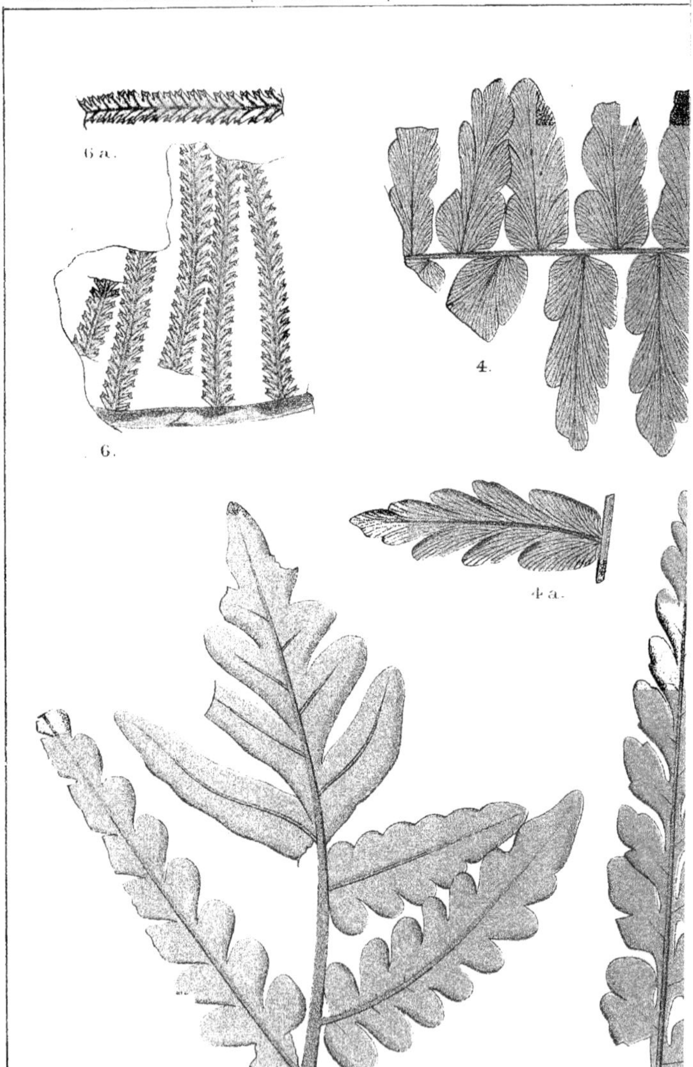

Fred. Earhart, del.

John S. Hart, Lith. Harrisburg, Pa.

1.

Fred. Earhart, del.

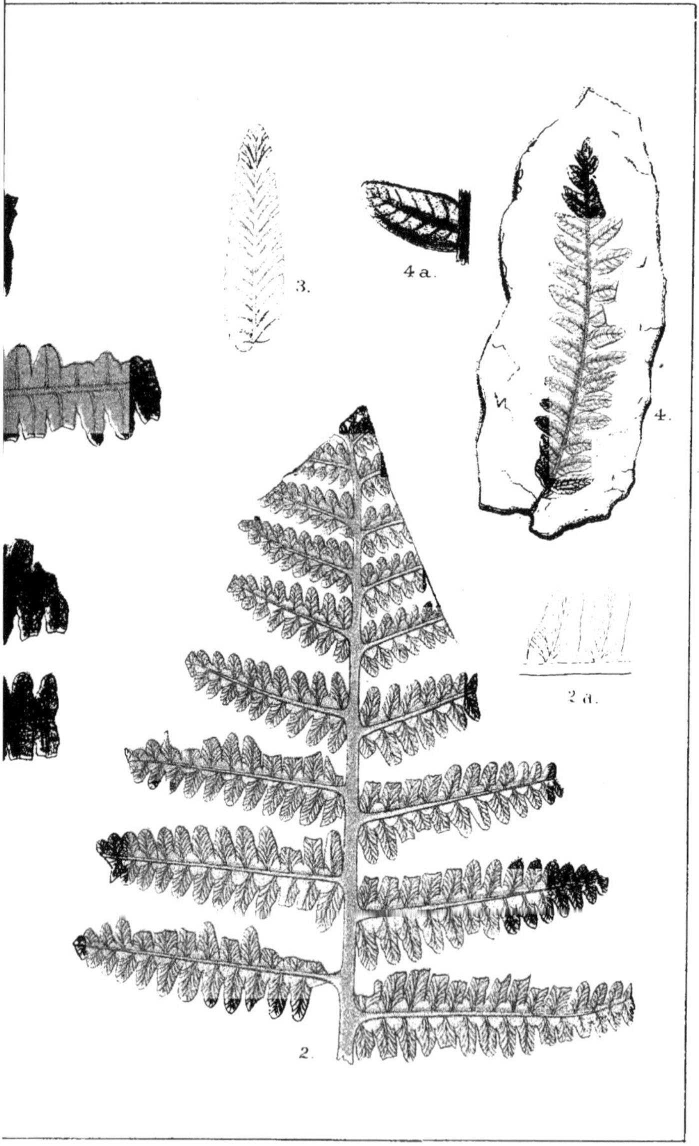

Lane S. Hart, Lith. Harrisburg, Pa

5.

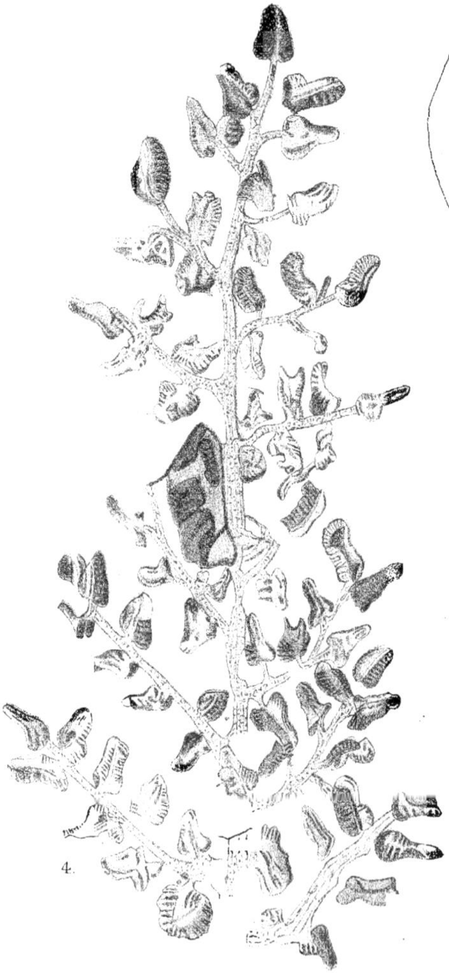

4.

Fred. Earhart, del.

3 a.

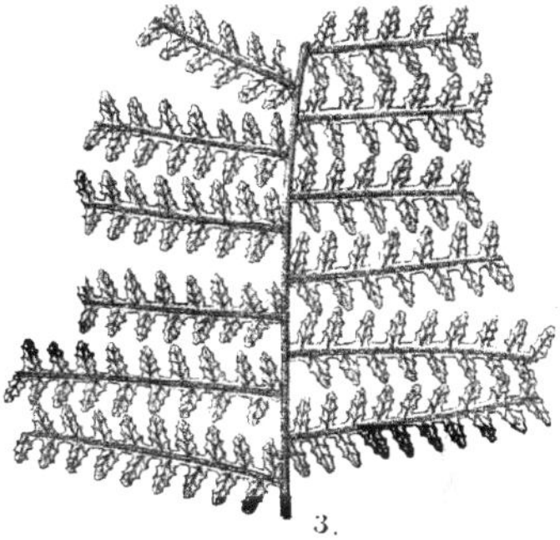

3.

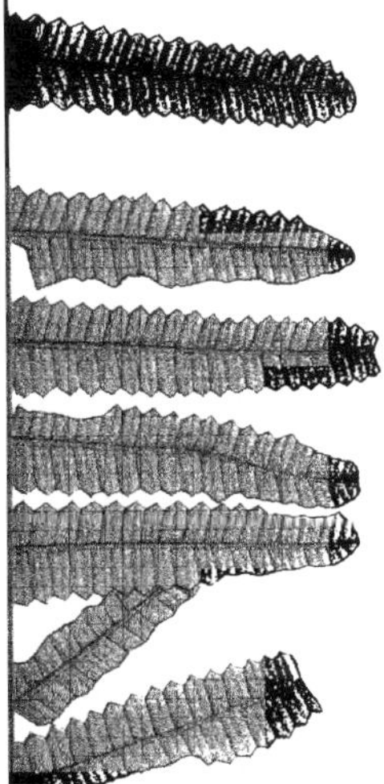

1.

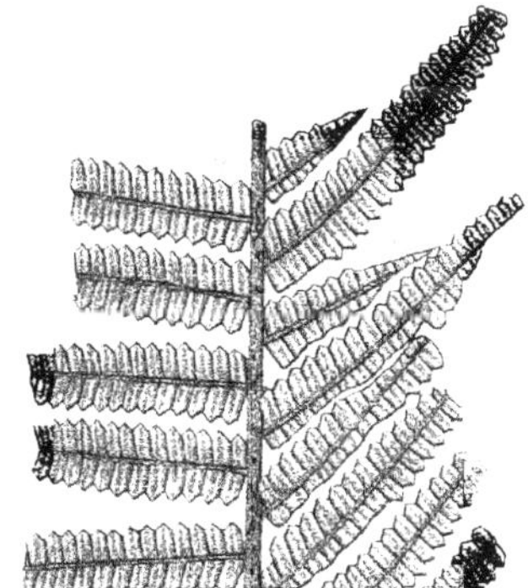

2.

Lane S. Hart, Lith. Harrisburg, Pa.

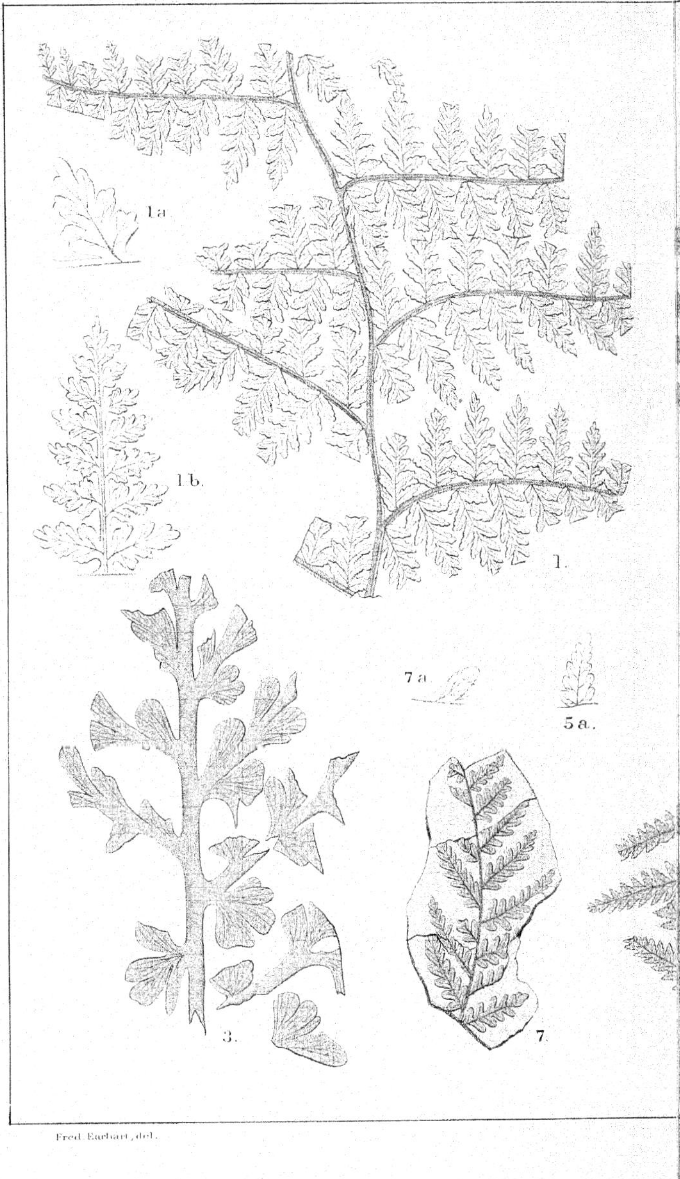

Fred. Earhart, del.

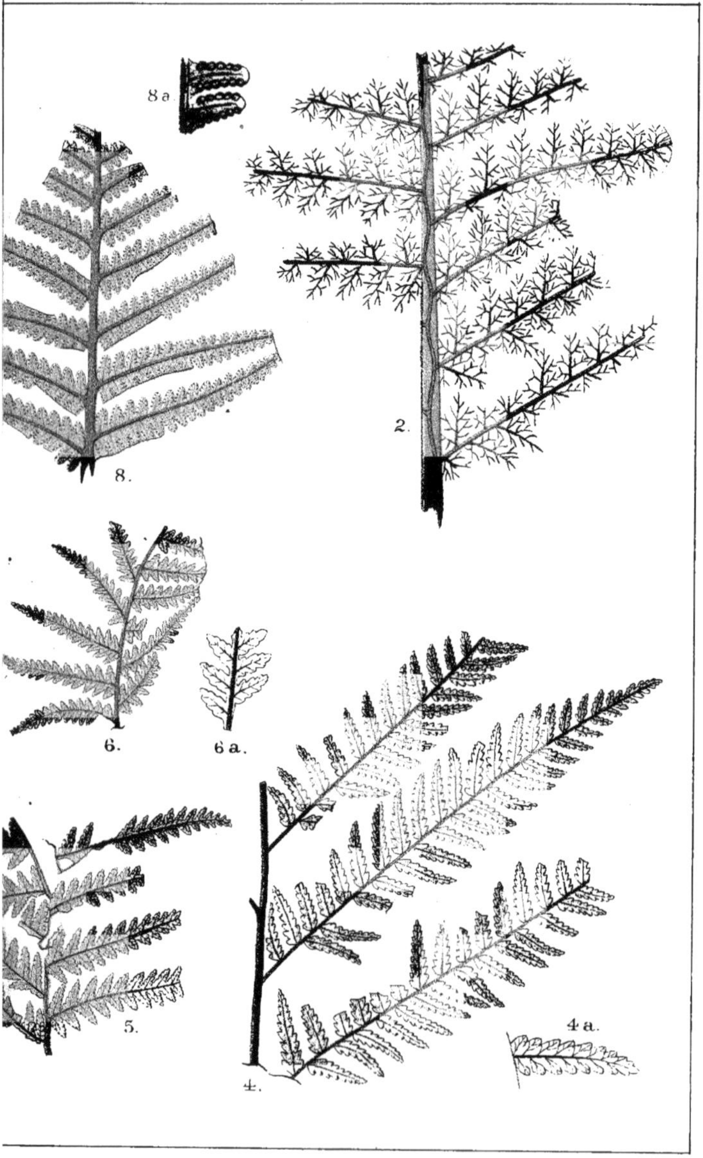

Lane S. Hart, Lith. Harrisburg, Pa.

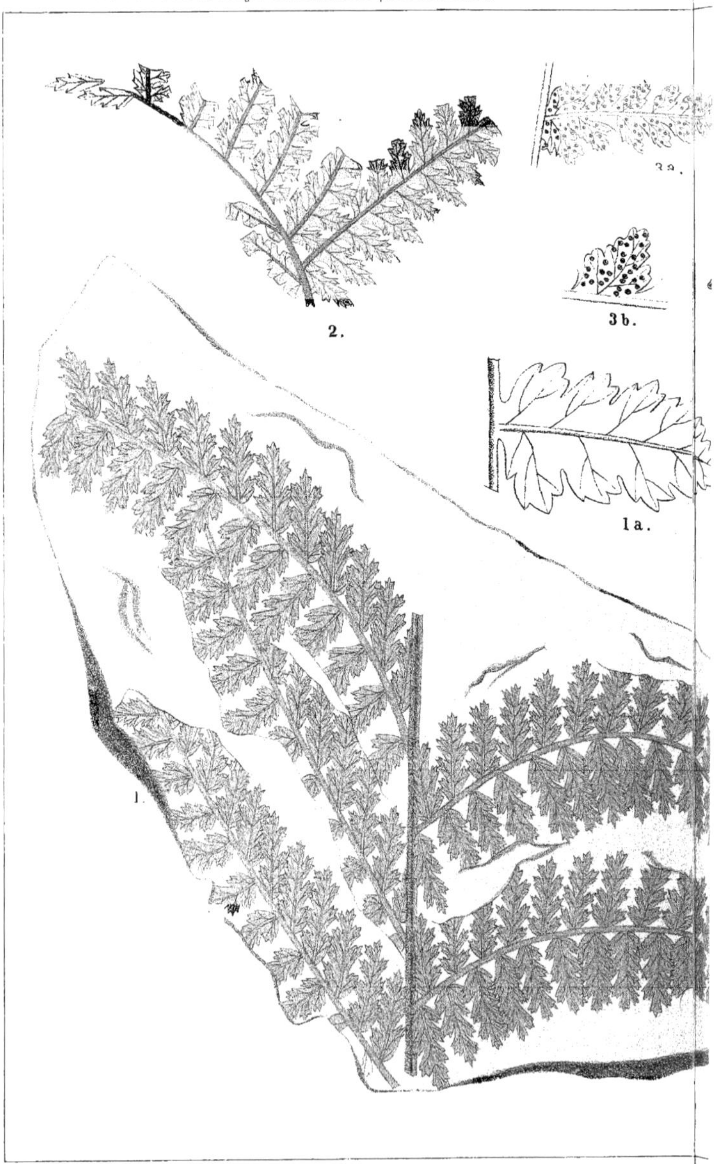

Ferd Earhart, del.

Lane S. Hart, Lith. Harrisburg, Pa.

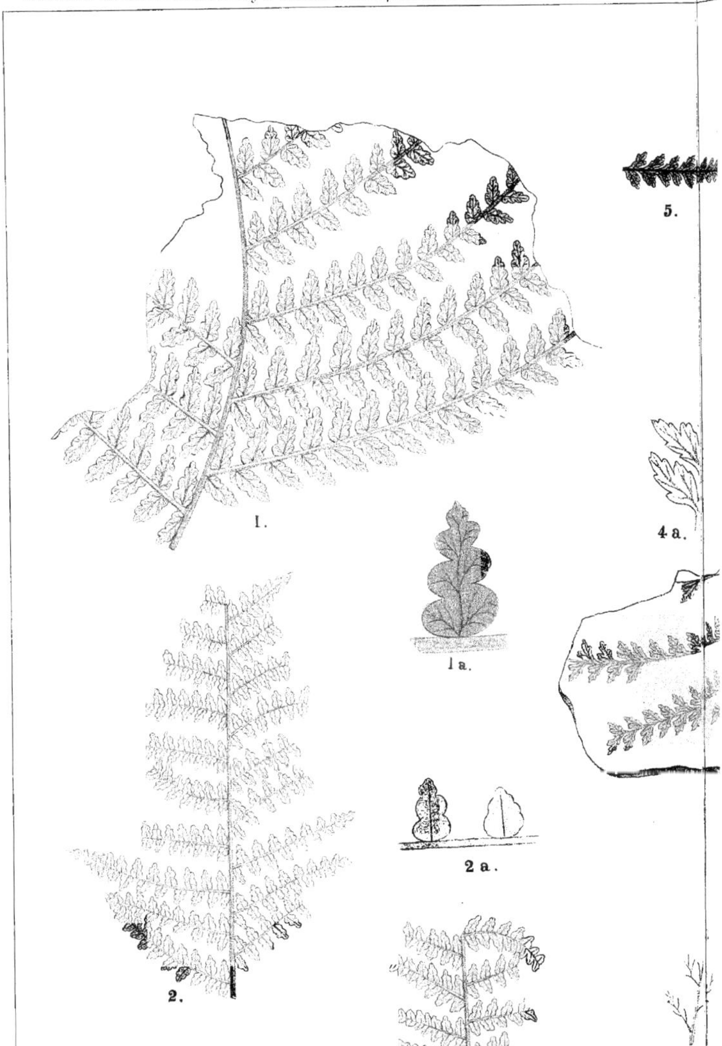

Fred. Earhart, del.

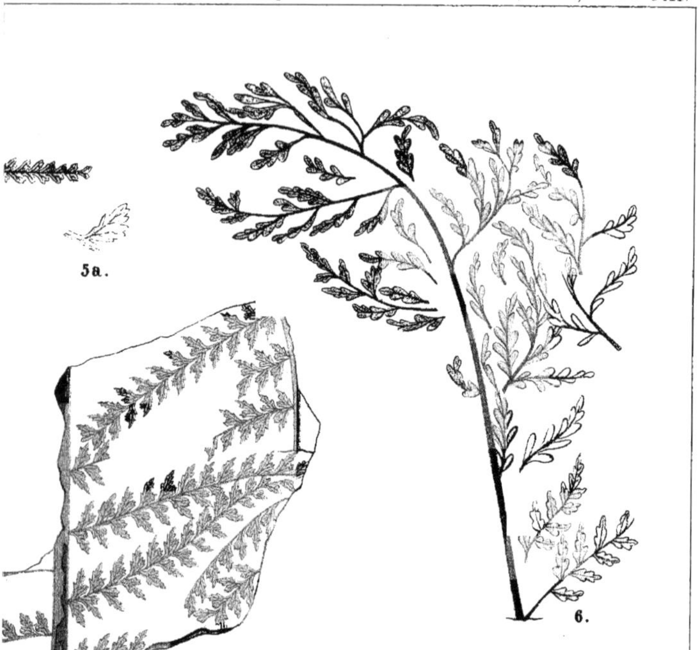

Lane S. Hart, Lith. Harrisburg, Pa.

Fred. Earhart, del.

Lane S. Hart, Lith. Harrisburg, Pa.

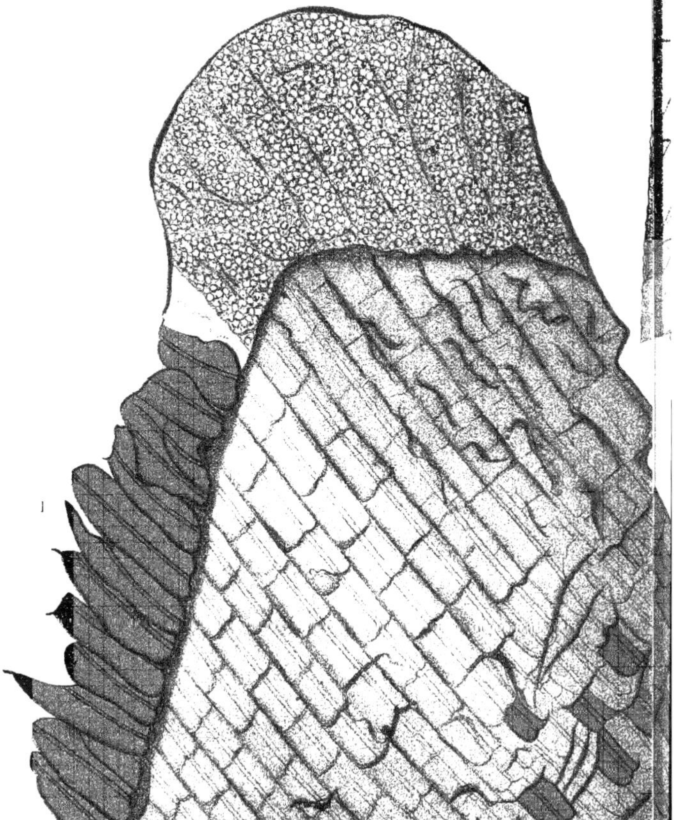

Fred. Earhart, del.

2. 3. 4.

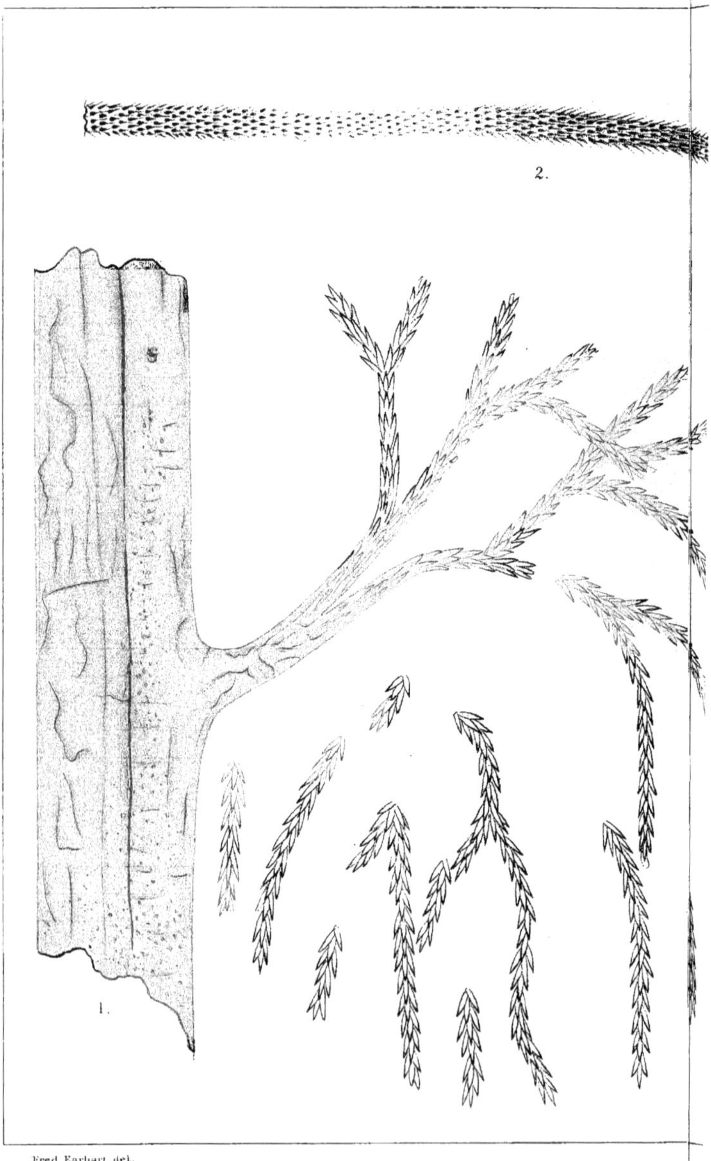

Fred. Earhart, del.

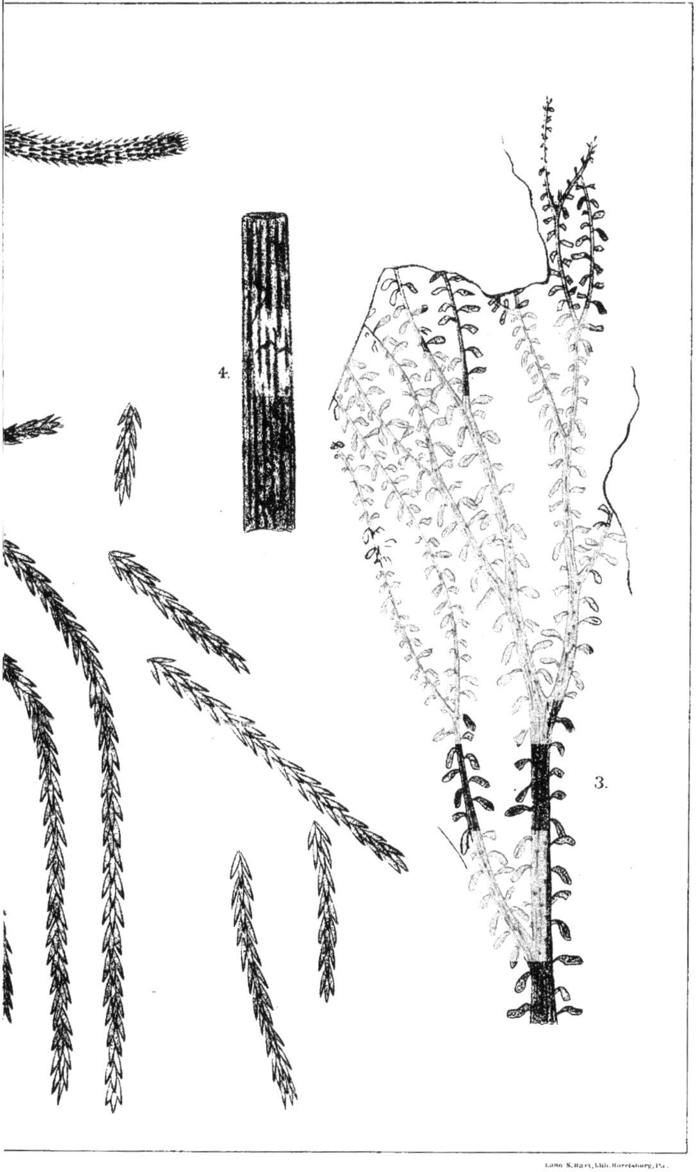

4.

3.

Lane S. Hart, Lith. Harrisburg, Pa.

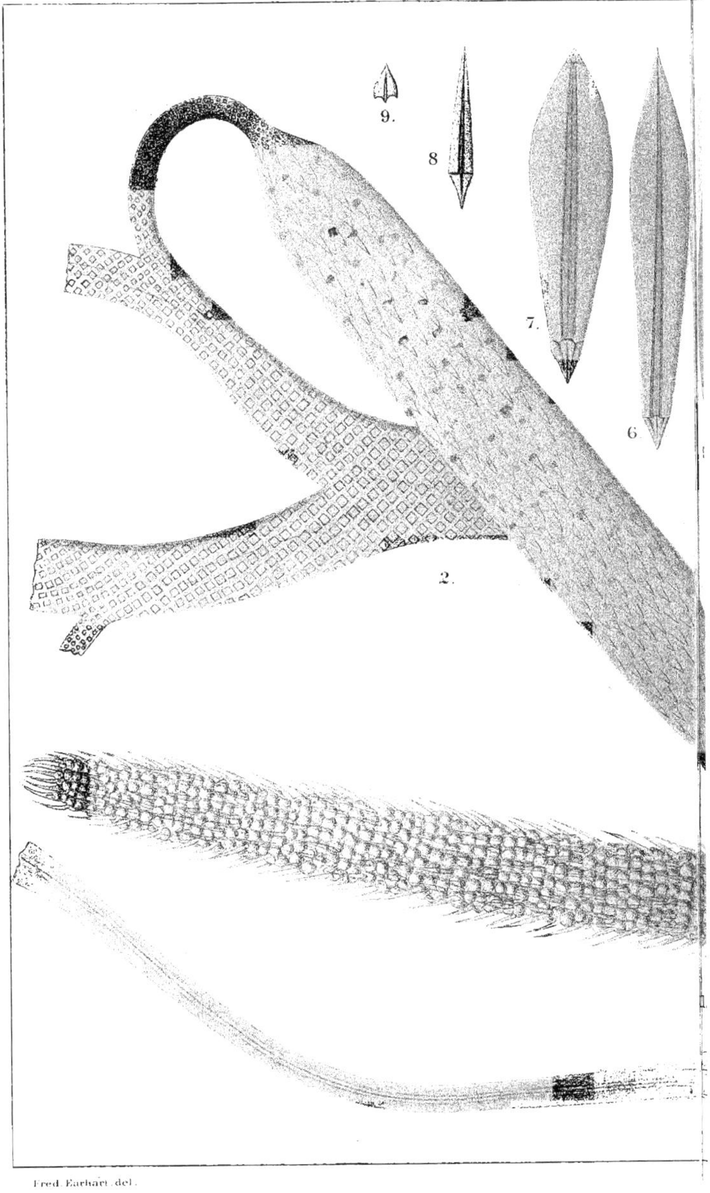

Fred. Earhart. del.

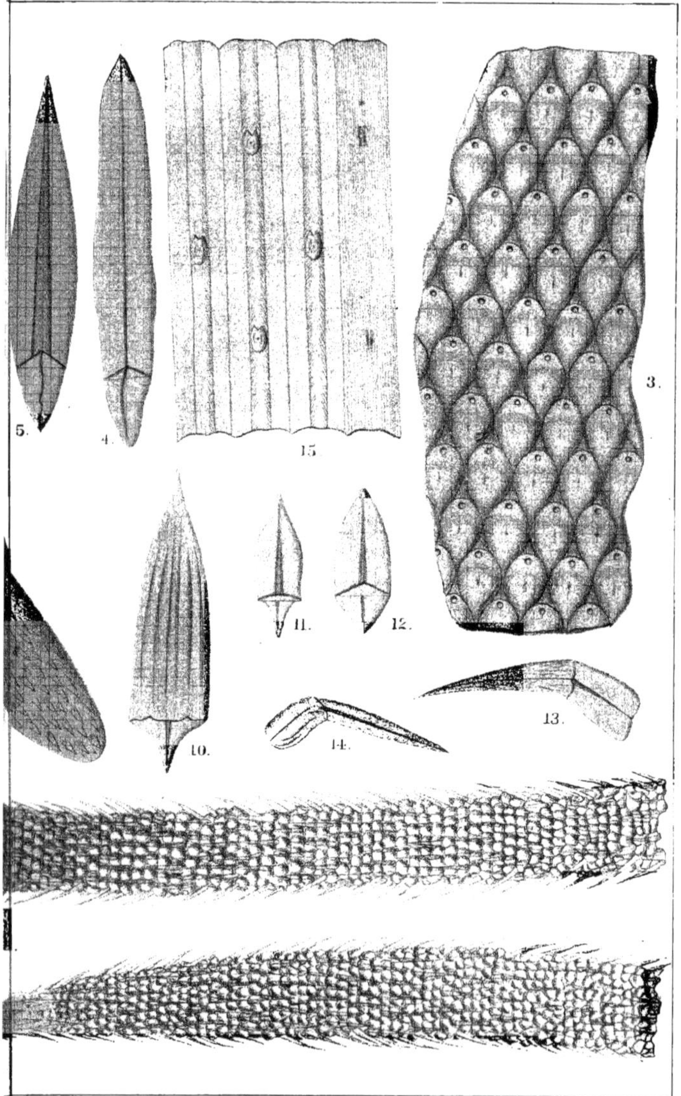

Lane S. Hart, Lith. Harrisburg, Pa.

Fred. Earhart, del.

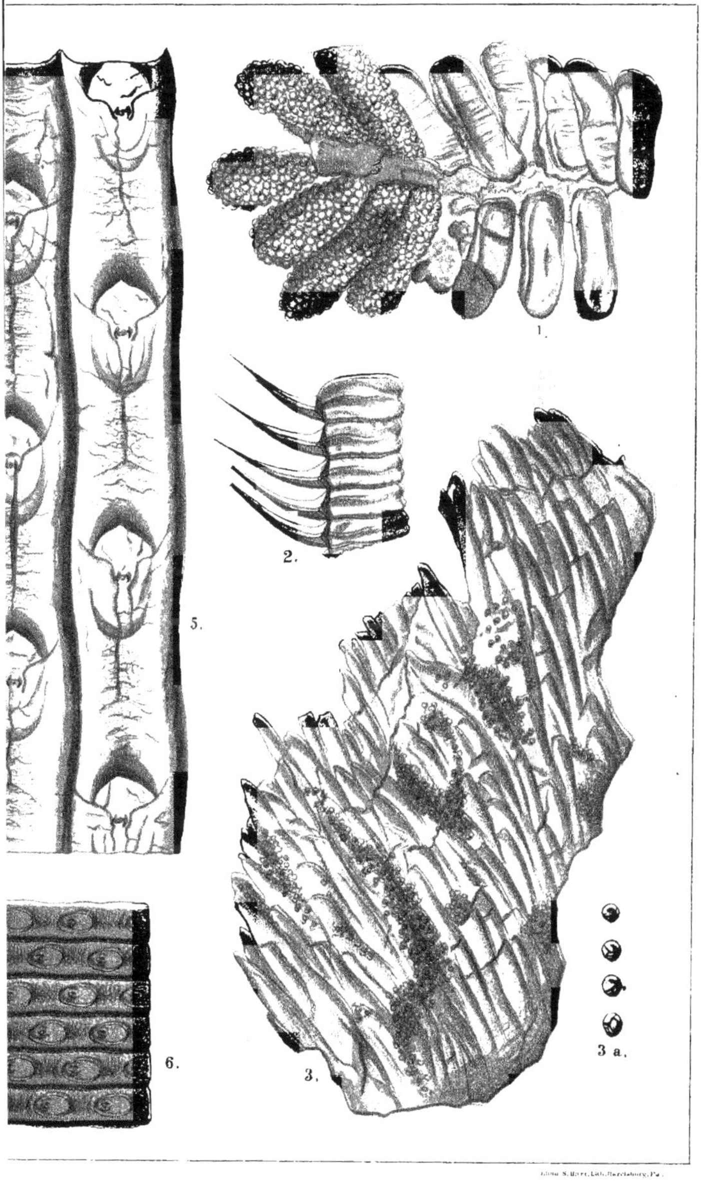

Geo. S. Harris, Lith. Harrisburg, Pa.

Second Geol. Sur. Pennsylvania, L. Lesquereux 1883.

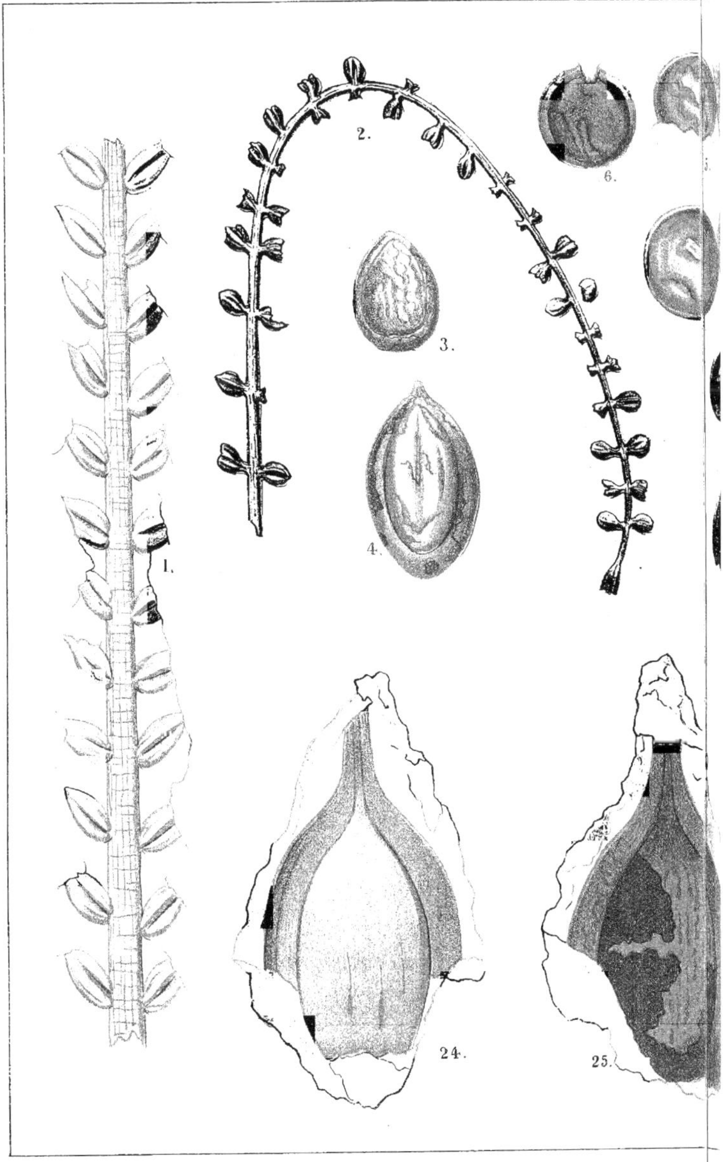

Fred. Earhart, del.

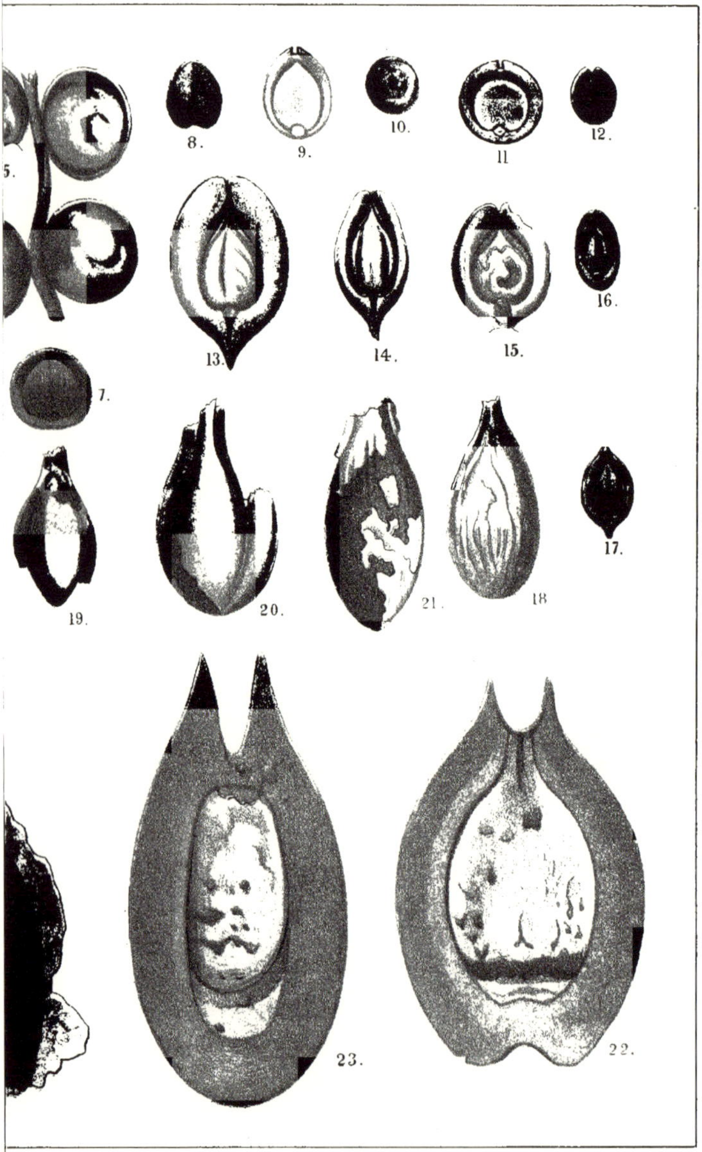

Louis N. Hart, Lith. Harrisburg, Pa.

Second Geol. Sur. Pennsylvania, L. Lesquereux 1883.

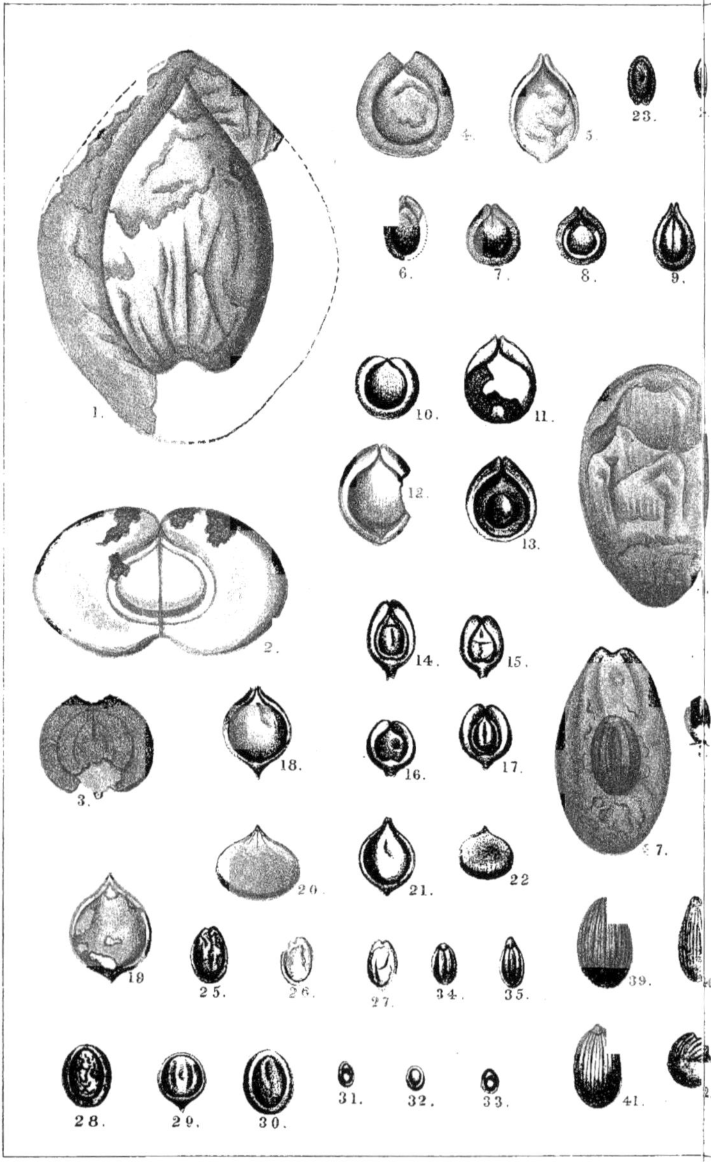

Fred. Earhart, del.

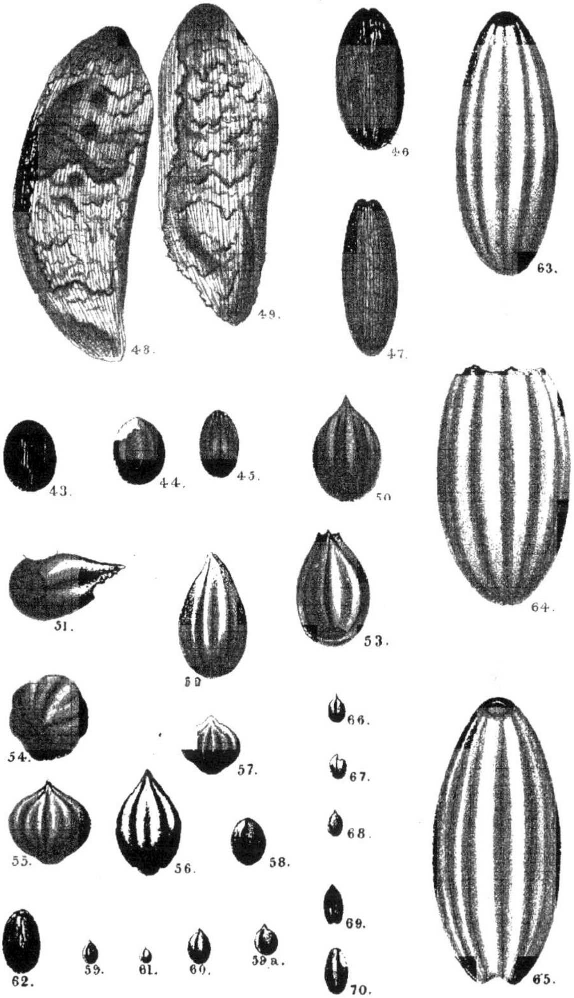

Lane S. Hart, Lith. Harrisburg, Pa.

Fred. Earhart, del.

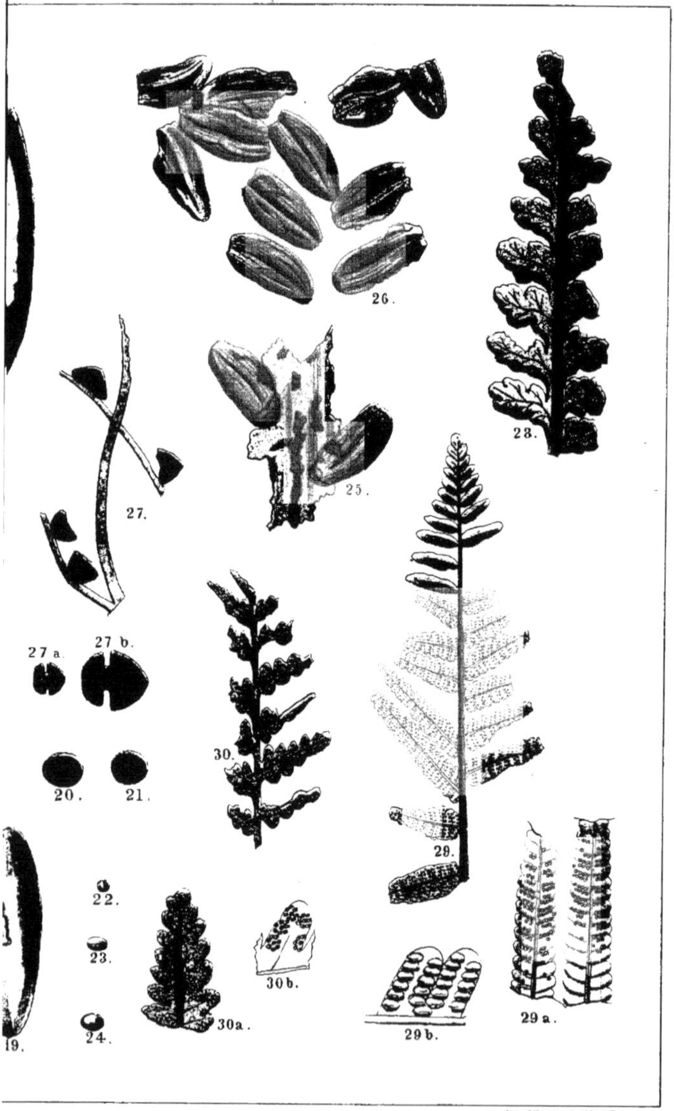

Lane S. Hart, Lith. Harrisburg, Pa.

Spotted something that isn't quite right?

Let us know and we'll fix it for future reprints.
And as a thank you for your diligence, we'll
send you a free Sothis book of your choice.

publisher@sothis.press

This page intentionally left blank

Made in United States
Orlando, FL
05 November 2026

90293405R00219